Friederike Franziska Lott

Wind Systems in the Dead Sea and Footprints in Seismic Records

Wissenschaftliche Berichte des Instituts für Meteorologie und Klimaforschung des Karlsruher Instituts für Technologie (KIT)
Band 70

Herausgeber: Prof. Dr. Ch. Kottmeier

Institut für Meteorologie und Klimaforschung
am Karlsruher Institut für Technologie (KIT)
Kaiserstr. 12, 76128 Karlsruhe

Eine Übersicht aller bisher in dieser Schriftenreihe
erschienenen Bände finden Sie am Ende des Buches.

Wind Systems in the Dead Sea and Footprints in Seismic Records

by
Friederike Franziska Lott

KIT Scientific Publishing

Dissertation, Karlsruher Institut für Technologie (KIT)
Fakultät für Physik, 2016
Referenten: Prof. Dr. Christoph Kottmeier, PD Dr. Joachim Ritter

Impressum

KIT Scientific Publishing

Karlsruher Institut für Technologie (KIT)
KIT Scientific Publishing
Straße am Forum 2
D-76131 Karlsruhe

KIT Scientific Publishing is a registered trademark of Karlsruhe Institute of Technology. Reprint using the book cover is not allowed.

www.ksp.kit.edu

Print on Demand 2017 – Gedruckt auf FSC-zertifiziertem Papier

ISSN 0179-5619
ISBN 978-3-7315-0596-9
DOI: 10.5445/KSP/1000060565

Abstract

Atmospheric processes, from microscale turbulence to severe storms on synoptic scale, have an impact on the continuous ground motion of the earth. Seismological recordings, such as the ground motion velocity, are affected by the ambient seismicity within a broad frequency band. To understand the influence of wind on seismological records, not only the genesis, the intensity and the spatial and temporal extent of meteorological systems must be considered. Investigations further require fine-meshed comparisons of meteorological and seismological measurements acquired at the same location.

In the scope of the HGF project DESERVE, a target-oriented experiment to study the influence of wind on ground motion was designed and conducted close to the Dead Sea in Jordan. The Dead Sea valley is subjected to unique atmospheric conditions, where various distinct local wind systems exist in summer and storms in the winter season are common. For the period from March 2014 to February 2015 a dedicated seismological array, consisting of 15 three-component short-period and broad-band stations, was installed near Madaba. It was complemented by a meteorological station providing three-component wind records with a 20 Hz sampling rate.

This thesis investigates the ground motion in relation to atmospheric motion processes. The observation parameter, recorded by seismometers, thereby is the ground motion velocity, in particular the power spectral density of ground motion velocity. By establishing a collocated, seismological and meteorological set of recordings, the scaling of the power spectral density of ground motion velocity with the parameters as the horizontal wind speed and wind direction is investigated.

Severe wind events manifest as distinct, wind-induced signals in spectrograms of seismic ground motion velocity. Thereby, the genesis of the wind field can be ascribed from the seismological records, exhibiting ocean microseisms from the Mediterranean Sea, microseisms from the Dead Sea, and local excitation of ground motion velocity. Examining the performance of seismological and meteorological records, the mean horizontal wind speed and the power spectral density of the ground motion velocity turn out as convenient parameters to quantify the impact of wind-induced shear on seismological records.
The power spectral density of ground motion velocity reveals a predominantly linear increase with the horizontal mean wind speed. Defining the ground motion susceptibility as a parameter of the increase with wind speed, a universal measure to estimate the vulnerability of seismic ground motion velocity to the excitation by wind is assessed. Spectra of power density of ground motion velocity, calculated for specific wind speed levels, demonstrate the broad-band character of wind-induced seismicity and prove that ground motion velocity is prone already to wind speed below $2\,\mathrm{m\,s^{-1}}$. Across plain, homogeneous terrain, the horizontal wind speed is considered to be an appropriate parameter to estimate the ground motion susceptibility. Complex topography, altering the horizontal wind flow, requires to consider additionally the wind shear, respectively the friction velocity of the horizontal wind field above the surface.
This thesis defines a measure for the impact of wind on seismic ground velocities and demonstrates the influence of specific wind speed levels on the power spectral density of seismological records in relation to the common New Low Noise Model (Peterson, 1993) for seismological recordings. The results reveal that wind on the level of a light breeze can have an observable impact on seismological records. When assessing the noise level of seismological recordings, therefore the impact of wind is a factor that should not be ignored.

Contents

1 Motivation

Seismology, as we know it today, is a relatively young science. Although the first pendulum seismometer, able to detect an earthquake, can be dated back to a Chinese philosopher in 132 A.D. (Dewey and Byerly, 1969), the evolution of seismology as a scientific discipline had to await both the theoretical background and the technical progress. The theory on elastic waves, built on Hooke's Law and later Poisson's Law (Lay and Wallace, 1995), is the premise for understanding the propagation of elastic waves in the earth. With technical progress, leading to an increase of instrument sensitivity, not only strong signals such as major earthquakes but also smaller, and more distant tectonic ruptures, and microseisms, originating from the earth's surface, could be recorded. Fusakichi Omori (1899) was one of the first seismologists who published convincing records of ocean microseisms, obtained with a specifically developed horizontal pendulum seismometer (Davison, 1924).

Microseisms, or seismic noise, is caused by vibrations of the earth's surface, mostly originating from the presence of humans, accompanied by traffic and industry, or from friction between air and solid surface, excited by wind as well as ocean waves. Thereby, obstacles, such as mountains, vegetations, buildings, and the oceans can reinforce the effects, vibrating and transmitting energy into the ground (Groos, 2010). Bonnefoy-Claudet et al. (2006) review the state of the art on the ambient seismic noise wave field embracing the discussion of sources and characteristics of the observations.
Coming along with the increasing sensitivity of seismometers, seismic noise has obtained significant importance and the demands on seismological recordings have grown. Today, ambient seismic noise can be considered under two

main aspects: on the one hand it can be the worst threat to seismological records, superposing and interfering desired signals from tectonic events. On the other hand, some approaches, such as seismic interferometry, take advantage of the ambient, uncorrelated nature of seismic noise in order to receive the impulse response of the propagation media (e.g. Wapenaar and Fokkema, 2006).

Wind-induced and anthropogenic noise both cause ambient ground vibrations over broad frequency bands (McNamara et al., 2009; Wilcock et al., 1999). Both types influence seismological records in a similar ways, however they are distinguishable by temporal occurrence. Anthropogenic noise, though subject to diurnal variations, may be treated as a site specific parameter that, once assessed, can be estimated for given weekdays and time of days (Groos and Ritter, 2009). Wind-induced noise presents a particular challenge, since it strongly varies with the temporally changing wind field. While the vulnerability of a seismological station to anthropogenic noise may be very site-specific and subjected to diurnal variations, the influence of wind needs to be assessed as a dynamic function of the local wind and the synoptic situation (Tanimoto and Artru-Lambin, 2007).
Wind affects the seismic ground motion velocity over a broad frequency band in ranges between millihertz and tens of hertz. Thereby, the specifications depend on the presence of potential obstacles and on whether the excitation has a local or distant origin (Withers et al., 1996; Saccorotti et al., 2011; Bromirski et al., 2005). In recent years, several studies investigated the impact of wind on ground vibrations by comparing seismological records to publicly available recordings of wind or meteorological reanalysis models. Holub et al. (2008, 2009) analysed the influence of the European storms Kyrill (January 2007) and Emma (February 2008) on seismological records. By comparing wind speed records to seismological records they found temporal correlations between periods of enhanced wind speed and increased microseisms. One step further, Ritter and Groos (2007) illustrated an example of the energy density of ground

motion velocity in the presence of storm Kyrill, demonstrating the excitation of ground motion velocity in different frequency domains.
Whereas these studies are limited to case studies on strong wind situations, other studies, such as Wilcock et al. (1999); Saccorotti et al. (2011), examine the dependency of the seismic noise level with respect to wind speed quantitatively. The majority of studies (e.g. Essen et al., 2003; Dahm et al., 2006) on wind-induced seismic noise deals with ocean microseisms in seismological records, caused by ocean waves hitting the sea floor and the coasts. Strong wind plays an indirect role, triggering ocean waves that swell with increasing wind speed (Pierson and Moskowitz, 1964). Hundreds of kilometres away from the coasts, ocean microseisms can be identified in seismological records, assigned to increased wind speed and to characteristics of ocean waves (Schulte-Pelkum et al., 2004; Zhang et al., 2009; Essen et al., 2003). Thereby, seismological records in frequencies well below 1 Hz are mainly affected (Friedrich et al., 1998).

Like oceans, obstacles at the land surface can enhance the influence of wind on seismological records. Saccorotti et al. (2011) found that in the vicinity of a wind farm the power spectral density of ground motion velocity dramatically increased under the influence of wind, compared to reference measurements in larger distances to the wind farm. There are only few studies on the influence of the local wind field on seismological records (e.g. Gerstoft et al., 2008; Withers et al., 1996; Wilcock et al., 1999; Dahm et al., 2005). Therefore, it has not been possible so far to develop a general empirical dependency of the vulnerability of seismological records to the influence of wind, as initiated in Thun (2011). In particular, there is a lack of studies on the affection of ground motion velocity to microscale wind events in the absence of amplifying obstacles. While ocean-generated microseisms are straightforward to identify in seismological records due to the distinct frequency band affected below 1 Hz, it is a greater challenge to deduct the influence of local obstacles, such as buildings or wind farms.

This thesis combines two main approaches from previous studies:

1) The investigation of case studies of storms with different genesis and analysing the influence of wind on ground motion velocity as wind-induced signals in order to investigate the origin of the sources

2) The examination of the systematic influence of wind for a continuous period

The Dead Sea valley, where distinct, local wind systems interact in summer and occasional storms in the winter seasons are common (Alpert et al., 2004a), provides convenient conditions to study the influence of wind on ground motion. The area is sparsely populated, offering an undeveloped environment with rare buildings and vegetation. From March 2014 to February 2015 a temporary seismological array, consisting of 15 broad-band and short-period stations, along with a meteorological tower, providing turbulent wind measurements, was installed in the Dead Sea valley in Jordan. By assembling meteorological and seismological records, wind-induced signals of ground motion velocity are assigned to microscale wind processes and mesoscale systems from distant sources such as storms of various genesis. Empirical seismic power spectral density spectra are developed, revealing the sensitivity of the ground motion velocity to the presence of wind. Finally, a systematic approach is established to describe the power spectral density of ground motion velocity as a function of the mean horizontal wind speed.

The synthesis will help to improve and refine the knowledge on wind-related noise in seismological records. The mean horizontal wind speed is discovered to be a feasible parameter to scale the forces at the interface between atmosphere and ground, suitable to assess the impact of wind on ground vibration. Bridging the gap between theoretical considerations and empirical findings, the present study is divided into three main parts.

Chapters 2 and 3 present an outline on meteorological, fluid mechanical and seismological considerations on applied forces between the atmosphere and the solid ground, followed by a brief description of the Dead Sea region. Chapters 4 and 5 illustrate the experiment and methods, specifically developed in this thesis and essentially necessary to understand considerations, calculations, and results. Finally, the results in this thesis are advanced by two approaches: case studies on severe wind events are investigated phenomenologically in Chapter 6, adding interpretation of GFS model analysis. The findings lead towards a qualitative understanding of wind-induced signals in seismological records, differentiating signals excited by various local and distant sources, including ocean microseisms. Building on this knowledge, chapter 7 develops a systematic scaling approach of power spectral density of ground motion velocity with horizontal wind speed: the ground motion susceptibility. Including considerations on shear stress at the interface between atmosphere and solid ground, the empirical dependency of power spectral density of ground motion velocity on wind speed is brought into context with theoretical implications.

It is not possible to prevent seismic noise of all kinds. However, it is possible to learn about the sources, the dependencies, as well as the triggers, and finally reveal noise characteristics. Though, it has been proven for more than a century that wind, directly or indirectly, is able to trigger vibration of the earth's surface, theoretical knowledge on applied contact force between atmosphere and solid ground has not yet been quantitatively brought into context with ground vibrations measured at one specific seismological receiver. This study is intended to evaluate a systematic scale for vulnerability of ground motion velocity to shear from the atmosphere, quantified in relation to wind speed.

2 Forces between Atmosphere and Ground

To understand how an atmospheric process such as the wind can affect seismological measurements, it is essential to give attention to the formation of seismological signals and on how the solid earth is influenced by tectonic ruptures and forces at its surface. The earth oscillates due to a broad band of causes, the strongest of these incidents originating from dynamic processes in the mantle of the earth. However, many sources for seismological signals arise from the surface of the earth or processes in farer distances outside, such as the tides, also exciting vibrations of the earth. To investigate the excitation of specific signals, such as caused by wind, it is necessary to describe the driving forces to understand the relation between recorded signals and signal causes.

2.1 Seismological Signals and Recordings

Unlike stationary measurements of meteorological parameters, such as wind speed or barometric pressure at a meteorological tower, measurements at seismological stations at or within the earth's surface can be described as omnidirectional, passive remote sensing methods. Ground motion velocity, in fact, is recorded in-situ, however the signals travel from all spatial directions from source to receiver, modulated and damped on its progression through the medium. Without any a priori assumptions on the specifications of the ground, it is not possible to determine the source and location, nor the physical description of the signal from one single recording. Recordings of tectonic signals have to undergo a series of processing steps before, in an under-determined problem, the source parameters, such as the Seismic Moment or the stress drop of a tectonic rupture can be determined with model assumptions for the earth.

Understanding the propagation and modulation of a signal in the earth before recorded at a receiver, is a requirement to interpret physical sources. The propagation of signals caused by vibrations of the earth's surface follows the same physical laws, however, the mechanisms are different from tectonic ruptures and therefore common tectonic source model assumptions fail (Section 2.1.2). For further reading on the seismological background, outlined in this section, Bormann (2012) can be recommended.

2.1.1 Interpretation of Tectonic Signals

Seismic waves are waves of energy, caused, for example by ruptures in the earth's crust, travelling through the earth, and recorded as seismological signals in the form of ground motion velocity by seismometers. Waves, propagating in a medium with dimensions significantly larger than the wave length itself, satisfy the WKB approximation[1] and can be described using ray theory by the eikonal equation (Lay and Wallace, 1995). By travelling through the medium, the waves encounter modulation, geometrical spreading, and attenuation due to specific properties of the leading medium. The loss of signal amplitude A due to geometrical spreading thereby depends on the divergence of the wave and the distance r to the source: $A \propto r^{-1}$ for body waves and $A \propto r^{-0.5}$ for surface waves. Therefore, signals recorded close to the source have high signal amplitudes, whereas far-distant signals have smaller amplitudes. Surface waves outrange body waves.

Earthquakes sources are localised sources, rupturing with finite duration. They radiate transient seismic signals decomposable into harmonic components whose phase relationships can be defined by the phase spectrum. The localisation of earthquakes, based on seismological signals and necessary to calculate earthquake source parameters, is a classic inverse problem. The aim is to find the location of the source and to determine the origin time. By

[1] Wentzel-Kramers-Brillouin (WKB) theory.

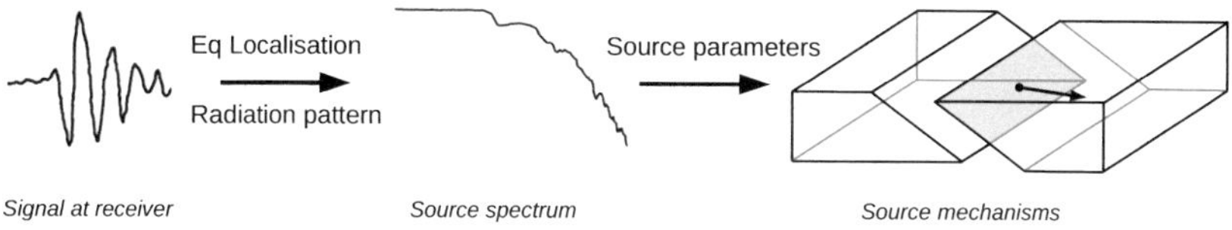

Figure 2.1: Schematic process of modelling earthquake source mechanisms, such as the moment tensor. Based on the seismological signal at the receiver, the earthquake can be localised. Deriving the wave radiation pattern and using a source model, the source spectrum can be estimated, leading to the source mechanics.

observing the differences in signal arrival times at multiple receivers a model of the underground that fits the recorded arrival times best can be found (Stein and Wysession, 2009).

To understand the physical parameters behind an earthquake, respectively the source parameters, not only a model for the medium in which the waves propagate, has to be determined, but a source model for the fault plane and the rupture process has to be established (cf. Figure 2.1). In the last century, the seismological community moved forward in determining the intensity of earthquakes not only by a function of epicentral distance and signal amplitude at the receiver, as used for the Local Magnitude (Richter, 1935). Instead, we can make assumptions about the source, that can be described as a very basic point source model, experiencing shear stress described by a force couple, or as more sophisticated models for large earthquakes (Haskell, 1964; Aki, 1967; Brune, 1970) depending on parameters derived from the spectra of the seismic signal.

The determination of earthquake intensity and stress release has a profound history and still, with enhancing prospects on the base of seismological records, models provide only estimations on actual physical occurrences, such as the release of stress during an earthquake. Only by knowing the hypocentral distances and assuming models for ground and signal source, it is possible to derive estimations on the stress release of a rupture based on seismological recordings of the far-field of the signal.

Seismic signals originating from earthquakes or blasts in the earth excite body waves and surface waves propagating with different medium specific seismic velocities, and with different frequencies. The frequency domain of a seismic signal highly depends on the hypocentral distance, and the extent of the earthquake fault plane. However, it ranges roughly between 0.1 Hz for teleseismic earthquakes and up to 50 Hz for local earthquakes.

2.1.2 Seismic Noise in Different Frequency Domains

Apart from sudden ruptures, such as related to earthquakes, the earth persistently vibrates. The sources are manifold, but many of them can be identified at the earth's surface. Other than distinct, locatable earthquake sources, ambient seismic noise is caused by a diversity of different, spatially distributed, sweeping, or continuous sources. As a non-stochastic process it can not be defined by a phase spectrum.
Seismic noise is constantly present and superposes all seismological records. Basically, it is the sum of all seismological signals that are not defined as prospected signals and due to its ambient character it is not divisible by its segments. Therefore, seismic noise can't be related to one single source, it is only possible to identify the aspects that contribute in certain traceable frequency bands. By observing the frequency domain in the form of spectra, noise characteristics can be analysed.
The new low noise model (NLNM), developed by Peterson (1993), describes the seismic background noise of the ground acceleration with respect to the frequency as an envelope of the lowest noise obtained from a vast network of seismological stations. Though there are more modern, statistical models by now (Castellaro and Mulargia, 2012), the NLNM is still a valid reference. The corresponding power spectral densities of seismic acceleration, velocity, and displacement added by Bormann (1998) as illustrated in Figure 2.2 describe the variations of the noise level with varying periods between 10^{-1} s and 10^{4} s. The power spectral density levels in periods longer than 10^{3} s shown in Figure 2.2

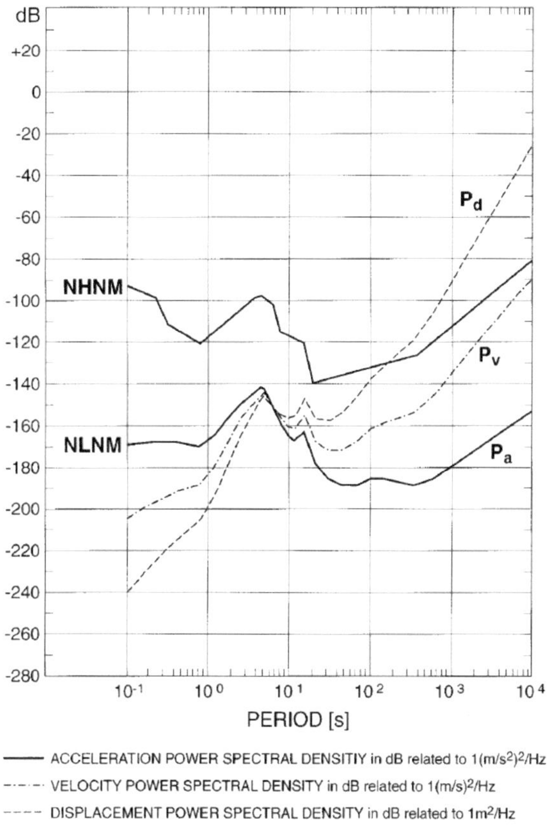

Figure 2.2: Envelope curves of the new global high and low noise models (NHNM and NLNM) developed by Peterson (1993). Acceleration power $\mathbf{P}_a$ in dB relative to $1\,m^2\,s^{-4}\,Hz^{-1}$, velocity power $\mathbf{P}_v$ in dB relative to $1\,m^2\,s^{-2}\,Hz^{-1}$, displacement power $\mathbf{P}_d$ in dB relative to $1\,m^2\,Hz^{-1}$. Illustration taken from Bormann (1998).

are not subjected to this thesis. Instead, key aspects are local noise in frequencies above 1 Hz, respectively periods below 1 s, and ocean microseisms, visible as peaks between 1 s and 30 s, attributed to ocean waves. Main contributions to the overall noise spectrum can be ascribed to anthropogenic influence, mainly traffic and industry, strong winds or barometric pressure changes, and ocean noise, coupled to the particular atmospheric situation. Noise can be classified by cause and dominates specific, broad frequency domains, depending on the cause. Note, that other than in Figure 2.2, results in this study are presented and discussed with respect to the frequency instead of the period.

Anthropogenic Noise

Anthropogenic influence is probably one of the most severe noise sources. A conglomerate of traffic, industry, and the general presence of humans causes ambient vibrations in a broad frequency band, most dominant in frequencies above 1 Hz, as for example investigated by McNamara and Buland (2004). With increasing distance between densely populated areas, such as cities, anthropogenic noise decreases. Due to the intense impact on seismological records, it is one of the highest priorities to select locations of seismological stations accordingly, reducing anthropogenic noise to a minimum. With increasing volume of traffic in densely populated areas, anthropogenic noise is becoming a more and more severe problem.
Anthropogenic noise varies during the time of day and day of week, depending on human activity. Apart from that, the noise level is a mostly specific, stable site effect at a seismological station. Therefore, the noise level due to anthropogenic influence is approximately rateable once the data quality is assessed.

Ocean Microseisms

Ocean microseisms can be assigned to the wave oscillations of ocean waves, dominant in power spectra of seismological records between 0.05 Hz and 0.5 Hz (cf. Figure 2.2). Ocean microseisms are permanently present, however intensified and modulated by strong winds, visible in seismological records all over the world, even far away from oceans. Other than anthropogenic noise, ocean microseisms take only narrow, distinct frequency bands, associated with the genesis of ocean waves. It is to distinguish between primary ocean microseisms and secondary, or double-frequency microseisms, as illuminated by

Longuet-Higgins (1950). Primary microseisms have its origin in shallow waters close to the coasts. Potential energy of the elevated barycentre of the water waves is converted into seismic energy by vertical pressure variations over the ocean ground, and by waves hitting the shore lines. Thereby, the dominant period in the seismic spectra corresponds to the period of the oscillating water waves. Secondary ocean microseisms derive from standing water waves by reflected, oncoming ocean waves of equal period, superpositioning and generating the standing gravity waves of the double frequency. Secondary microseisms are more dominant in the spectra (Bromirski et al., 2005), whereas the dominating particle motion is of Rayleigh-wave type. More elaborate descriptions of the mechanisms of ocean microseisms are subject to Friedrich et al. (1998); Cessaro (1994).

Noise Related to Atmospheric Effects

Noise, related to wind, and caused by the friction of air masses at the earth's surface, can have similar effects on seismological records as anthropogenic noise. Especially strong wind can have severe effects on seismological data quality, as investigated in several studies from Holub et al. (2008, 2009); Mucciarelli et al. (2005); Ritter and Groos (2007). However, wind-induced seismic signals affect an even broader frequency band as wind can excite obstacles such as large water bodies. Wind can amplify the ocean microseisms additionally (Zhang et al., 2009) by exciting strong waves in the seas, therefore affecting also narrow frequency bands down to below 0.1 Hz. Further, changes of barometric pressure can change the soil surcharge, deforming the earth's surface and affecting the very low frequencies (Widmer, 1995; Zürn et al., 2007; Beauduin et al., 1996), apparent in Figure 2.2 as high noise level in periods longer than 10^3 s.

The main problem with wind related noise is that it is highly variable, scaling with the strength of wind (Moskowitz, 1964), and unlike anthropogenic noise not generally assessable for a specific site. Further, as apparent from this study, wind seems to amplify specific narrow frequency bands individual for each site. All noise sources have in common that it is not possible to eliminate their influence. Whereas it is possible to minimize the influence of anthropogenic noise by choosing an appropriate site, it is harder to minimize the influence of wind on seismological records. However, all noise sources are located at the earth's surface. Deploying the seismological stations inside boreholes in the earth significantly improves the signal-to-noise ratios with increasing distance to the earth's surface, as investigated by Withers et al. (1996), and apparent by the amplitude decay of seismic signals. Due to the high costs, only few seismological stations can be deployed deep in boreholes, minimizing, however not eliminating, the influence of seismic noise from the surface. Most permanent seismological stations, such as station GHAJ described in Chapter 4.3 are deployed a few metres below the surface. Due to the shortest distance of less than one metre to the surface, temporary seismological stations, such as the DES array (Chapter 4.1) are most prone to seismic noise. Because of the vicinity of most seismological stations to the surface strong noise is an issue that has to be dealt with (Trnkoczy et al., 2012).

The definition of the term *noise* in seismology, and as it is used in this chapter so far, normally conforms with ambient, diffuse signals, superposing distinct, desired signals, such as earthquakes in seismological records. However, the definition depends on the observer. Eventually, it depends on which kind of signals is to be investigated. Signals from other sources as desired, thereby jam the investigated signals and worsen the data quality.
This study varies in the definition of wind related impact on seismic ground motion velocity. As the influence of wind can be observed quantitatively over broad frequency bands in spectra of ground motion velocity, usually related to as noise level impacting the data quality, it is also investigated as time-variant parameter in the seismological records and therefore referred to as wind-related signals. With its strong, time dependent fluctuations in intensity, variable frequency content, and different durations, it exhibits certain characteristics of

seismic signals, similar to seismic tremor (Nadeau and Dolenc, 2005), and therefore is referred to as *signal* when investigated in the time domain.
Still, other than common tectonic signals, the source of wind-related signals can not be located, as it can not be approximated as finite source recorded in the far-field. Common assumptions for earthquake sources become invalid. Therefore, it is not possible to estimate the forces that cause the wind-related signals with common approaches.

2.1.3 Design of Electromagnetic Seismometers

To assess and interpret seismic signals it is essential to give some attention to the mode of operation of seismometers, as summarized for example in Asch (2009). Seismometers in this work measure ground motion velocity in three spatial directions: one vertical and two horizontal components. Each of the components is recorded using its own sensor inside the seismometer oriented in equivalent directions. To simplify matters, the following description will only consider one sensor.
Inside the frame of the seismometer a small mass is attached to a spring connected to the seismometer frame. As the ground beneath the seismometer accelerates, the seismometer frame moves and accelerates with the ground. However, the mass attached to the spring stays in inertia. As a consequence, mass and seismometer frame experience relative movement. Modern electromagnetic seismometers determine relative displacement or velocity using electromagnetic velocity and force transducers. Triggered by ground acceleration, a coil attached to the mass moves in the field of a permanent magnet fixed to the seismometer frame. The motion induces a voltage in the coil. The current in the coil produces a force, creating a force transducer whose responsiveness can be equated with the responsiveness of a velocity transducer. Geophones and electromagnetic seismometers additionally built in a feed back force transducer to oppose the motion and damp the mechanical oscillation of the sensor. Hence, seismometers don't record the initial ground acceleration. They mea-

sure the relative movement of an inertial mass to the seismometer frame, and sense the motion via electromagnetic induction. Seismometers convert the initial input signal $f(t)$, that is a function of the time t, with a displacement transducer and integrator to an electric output signal $g(t)$ that corresponds to the ground motion velocity. However, since a direct sensing of the input signal is not possible, a seismometer has specific influence on the output signal that, if known, can be segregated from the output signal to get a better estimation of the input signal $f(t)$, as described in Chapter 5.1.2. The following section provides a brief approach on the mathematical process of signal recording necessary to understand the processing routine to retrieve the input signal from raw data.

Mathematical Description of Seismic Signals and Recordings

To work with seismological raw data it is essential to consider the mathematics of the signal processing inside a seismometer. As described in Chapter 5.1 the original recordings of seismological raw data do not equal the factual parameter of the ground motion velocity in physical units. The recordings rather represent an electronic signal modulated by the seismometer itself that needs to be calculated back to receive a representation of the original signal.

Mathematically, a seismometer can be described as a linear time invariant (LTI) system, acting as a filter to the original signal. To understand the essential data processing applied to seismological records in this thesis the process of signal acquisition is derived in more detail in the appendix in A.1. A more detailed discussion can be found in Scherbaum and Schmidtke (2007) and Wielandt (2012) who deduce and describe signal processing routines in seismology in great detail.

2.2 Influence of Wind on the Earth's Surface

When interpreting signals in seismological records it is important to consider the type of signal source in order to interpret the data. Other than distant tectonic signals or blasts from inside the earth wind acting on the earth's surface can not be approximated as point sources. Instead, wind acts as air flow over a surface of extensive size. When assessing the contact forces thereby applied at the earth's surface associated to wind, the driving forces of wind and meteorological scales need to be considered.

2.2.1 Driving Forces of Wind

Wind is a direct consequence of pressure gradients in the atmosphere causing the air to seek equilibrium of forces. The forces acting can be described by the Navier-Stokes Equation, considering a pressure term, Coriolis force, gravity, and a friction term.

On synoptic scale, typically between 200 km and 5000 km (cf. Figure 2.3), wind can be estimated by the horizontal air flow of the geostrophic wind, neglecting friction forces and assuming constant pressure at given height. However, when assessing the complete picture, friction increases with decreasing height above the surface. The influence of friction from the surface deviates the true wind from the geostrophic balance from parallel to the isobars towards low pressure areas. Over the seas the deviation ranges approximately between 10° and 20°. Over land the friction term is stronger, deviating the wind by about 20° to 40°.

On local and regional scales, the description of wind becomes even more sophisticated. Friction does not only play an important role in the Planetary boundary layer (PBL), but the topography has to be considered. Changing topography, by valleys and hills or mountains, causes obstacles for air flow, adding enforced vertical movements contributing to the balance.

For the Dead Sea region the influence of the deep depression of the Dead Sea valley, the presence of the Dead Sea itself and the Mediterranean Sea as water bodies, can not be neglected when discussing wind and its signature in seismic records. In fact, wind is strongly driven by local attributes.

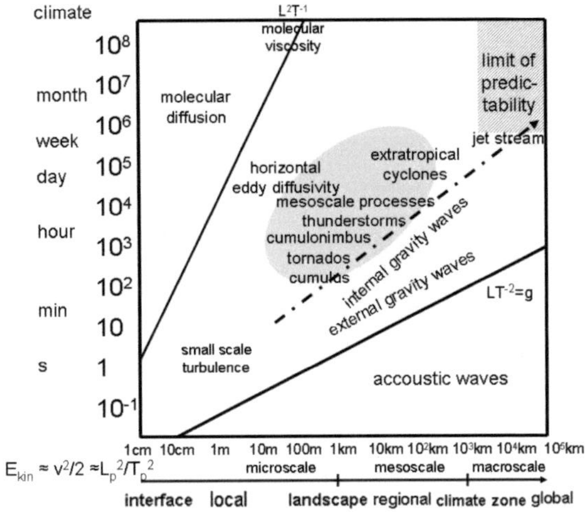

Figure 2.3: Meteorological time and spatial scales of motion, defined according to Orlanski (1975). From: Mölders and Kramm (2014)

Sea and Land Breeze

The interaction of land and sea breeze is controlled by temperature gradients originating in different thermal properties of water and land surface. Depending on the solar insolation, it varies for days and night times. Thereby, the temperature of the water body remains approximately constant, whereas the land surface temperature fluctuates faster, resulting in warmer temperature during day times and, in comparison to the water body, colder temperatures at night. Coming along with the temperature gradient, warm temperatures result in an uplift of wind, leaving low pressure areas at the surface, with pressure gradients towards the colder areas. Thus, at night, wind is directed from the colder land surface towards the warmer sea surface. During day time, the situation is opposed (Kraus, 2004).

Influence of Topography

In the presence of topography, the description of air flow, controlled by temperature and pressure differences, becomes manifold. In addition to anabatic

and katabatic winds, slope wind and channelled flow interact and need to be considered. Raupach and Finnigan (1997) outline the mechanisms and based on present investigations they assess the specific contributions.
Basically, like for land and sea wind circulation, wind circulation over rugged terrain can be explained by the same diurnal thermal processes of locally different rates of heating processes and resulting pressure differences. Adding anabatic and katabatic ascending and descending air masses leads to diurnally circulations of wind in the valley. Additional channelling of wind along the valley can be expected to be an important process for the Dead Sea valley as well, given the fact, that channelling effects even in the broad, shallow Rhine valley can be observed as well (Kalthoff and Vogel, 1992).
Important to consider in terms of local wind circulation is not only the direction of flow but the change of wind speed induced by varying shear at the surface. As discussed by Durran (1990) and Egger (1990), downslope winds, especially in the presence of stable stratification, may decelerate, whereas upslope wind can develop high speed.
Channelled wind flowing along valleys, can increase speed as well, while inside the valley, and decrease at the outflow. Thereby, it is a controversial issue, whether the valley acts similar to a Venturi tube, inside which the velocity of the fluid increases while the static pressure decreases due to a narrowing flow diameter, as stated by Lesouëf et al. (2011). Other studies, such as Koletsis et al. (2009) and Sharp (2002) claim that the highest wind speeds occur at the outflow of valleys as a result of strong pressure gradients.

2.2.2 Wind Shear in the Surface Layer

When investigating friction due to air flow above the surface of the earth, it is necessary to consider wind shear. In the free atmosphere, we can neglect friction and therefore have no vertical wind shear. In the PBL, however, wind is subject to shear, and with decreasing height above the surface, the horizontal wind speed decreases. The wind shear thereby determines the decrease of horizontal wind speed with decreasing height, and strongly depends on the stratification of the PBL (Figure 2.4). The lowest layer within the PBL, the Prandtl layer, is defined depending on the wind shear profile. It reaches from the surface up to where under neutral conditions approximately 80% of the wind speed above the PBL is detected (Stull, 1988). This often results in a thickness of the Prandtl layer of about 100 m, depending on the stratification. In the Prandtl layer the wind shear is highest, while wind speed decreases down to 0 m s^{-1} at the surface.

Wind speed measurements, used in this study, are provided by meteorological stations located well inside the Prandtl layer. While the driving forces above play a role for the genesis of the processes at the surface, the wind shear in the Prandtl layer is predominantly controlled by the mean horizontal wind speed and roughness length, respectively the topography. For investigations of friction forces and shear stress at the earth's surface, processes in the Prandtl layer are crucial.

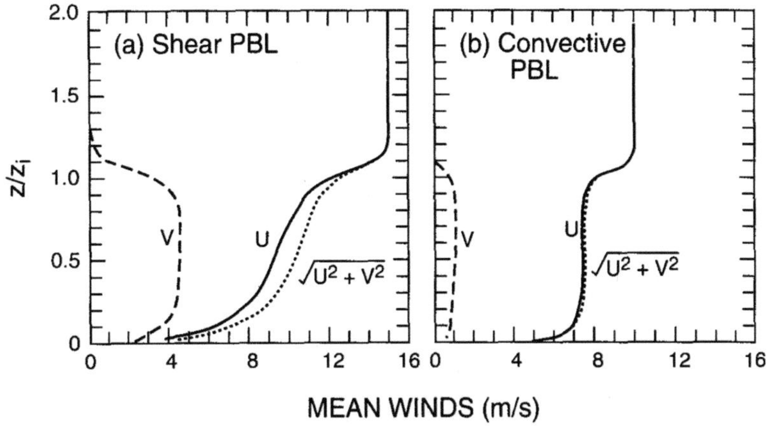

Figure 2.4: Schematic mean wind speed profiles in a shear dominated Planetary boundary layer (PBL) (a) and in a convective PBL (b). Horizontal mean wind speed: $\sqrt{u^2+v^2}$. z: Elevation above surface, z_i: PBL height. Taken from Moeng and Sullivan (1994).

2.2.3 Equilibrium of Forces between Atmosphere and Ground

The motion of viscous fluids can be described by the Navier-Stokes equations, built on the Cauchy momentum equation. To approximate solutions of the Navier-Stokes equations in meteorology, the model of the Reynolds-decomposition is common. The turbulent flow for a stationary, incompressible Newtonian fluid can be decomposed into (i.e. Stull, 1988):

$$u_i = \overline{u_i} + u_i'. \tag{2.1}$$

The turbulent description of wind speed u_i thereby equals the sum of its time average $\overline{u}_i$ and its perturbations u_i'. The Reynolds decomposition, applied to the Navier-Stokes equations, is an established approach to approximate turbulent fluxes. The approach results in the Reynolds-averaged Navier-Stokes equations. The notation in Etling (2008):

$$\underbrace{\frac{\partial \overline{u}_i}{\partial t} + \overline{u}_j \frac{\partial \overline{u}_i}{\partial x_j}}_{\text{I}} = \underbrace{-\varepsilon_{ijk} f_k \overline{u}_j - \frac{\partial \overline{\Phi}}{\partial x_i}}_{\text{II}} \underbrace{- \frac{1}{\overline{\rho}} \frac{\partial \overline{p}}{\partial x_i} + \nu \frac{\partial^2 \overline{u}_i}{\partial x_j^2} - \frac{\partial \overline{u_i' u_j'}}{\partial x_j}}_{\text{III}} \tag{2.2}$$

describes the floating mean of the turbulent impulse fluxes of an incompressible fluid without turbulent pressure perturbations. The inertial force (I) is depicted by the variation of the mean flow and the advection of the mean flow. It equals the sum of volume specific forces including body forces (II) and contact forces (III). The latter owe to isotropic stress, depending on the mean pressure field $\overline{p}$, viscous stresses, depending on the dynamic viscosity ν, and Reynolds stress $-\rho\overline{u_i' u_j'}$ evolving from the Reynolds decomposition, describing the turbulent fluctuation of the velocity field.

Neglecting internal body forces f_j and assuming a horizontal air flow across a surface, the geopotential Φ can be considered static with height. The term of the body forces (II) vanishes and the effecting forces in Equation 2.2 are reduced to contact forces (III):

$$\frac{\partial \overline{u}_i}{\partial t} + \overline{u}_j \frac{\partial \overline{u}_i}{\partial x_j} = \underbrace{-\frac{1}{\overline{\rho}} \frac{\partial \overline{p}}{\partial x_i}}_{\text{IIIa}} + \underbrace{\nu \frac{\partial^2 \overline{u}_i}{\partial x_j^2}}_{\text{IIIb}} - \underbrace{\frac{\partial \overline{u'_i u'_j}}{\partial x_j}}_{\text{IIIc}} . \tag{2.3}$$

Due to the low kinematic viscosity of air[2] (Stull, 1988) the viscous shear in the boundary layer is small compared to the Reynolds stress: IIIb $\ll$ IIIc. Therefore, it can be neglected here:

$$\frac{\partial \overline{u}_i}{\partial t} + \overline{u}_j \frac{\partial \overline{u}_i}{\partial x_j} \approx -\frac{1}{\overline{\rho}} \frac{\partial \overline{p}}{\partial x_i} - \frac{\partial \overline{u'_i u'_j}}{\partial x_j} . \tag{2.4}$$

Now the effective forces are described by the sum of shear forces, reduced to the Reynolds term, and dynamic pressure changes. Considering air flow across plain terrain, the horizontal flow is not subjected to dynamic pressure changes. It can be approximated by the wind shear, respectively the description of the Reynolds term alone. For more complex terrain with obstacles or elevations, however, dynamic pressure changes must be accounted (Chapter 7.2).

In the Prandtl layer the horizontal wind, sketched as roughly logarithmic profile in Figure 2.5, experiences strong shear due to the friction at the surface, where the wind speed is $0\,\mathrm{m\,s^{-1}}$. Based on assumptions of specific turbulent conditions, such as the gradient of the horizontal wind speed change with height $\frac{\mathrm{d}\overline{u}}{\mathrm{d}z}$, the Kármán constant (De Karman and Howarth, 1938), the surface roughness, and the stability, parametrized by the Monin-Obukhov parameter (Foken, 2006), it is possible to estimate the friction velocity from the mean horizontal wind speed.

Provided turbulent measurements of the three-dimensional wind field above the roughness length (cf. meteorological measurements in Chapter 4.2), however,

[2] For a typical mean wind shear in the atmospheric surface layers: the kinematic viscosity $\nu = 1.4607\text{E-}05\,\mathrm{m^2 s^{-1}}$. The resulting viscous shear stress is $7.304\text{E-}06\,\mathrm{m^2 s^{-2}}$

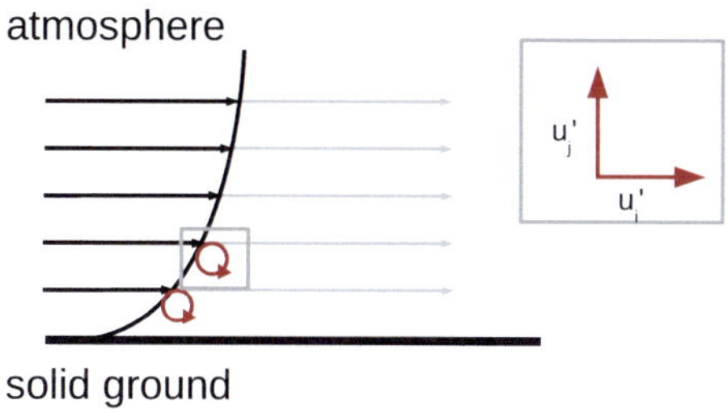

Figure 2.5: Wind shear caused by friction in the Prandtl layer above plane, solid ground. Black arrows indicate horizontal wind, decelerated with increasing proximity to the ground due to friction. Grey arrows indicate wind trajectories. Red circular arrows represent resulting turbulence, described by u_i' and u_j'.

the theoretical considerations in Equation 2.2 lead to more accurate descriptions. Reducing the shear stress resulting from the horizontal wind flow as the Reynolds term (Equation 2.3,IIIc), it is possible to derive the shear stress τ:

$$\tau = -\rho \cdot \overline{u_i' u_j'}, \tag{2.5}$$

whereas

$$u_* = \sqrt{\overline{u_i' u_j'}} \tag{2.6}$$

is defined as the friction velocity u_*. Thereby, ρ represents the air density, u_i' the deviation of the horizontal wind $\overline{u}_i$ from the mean horizontal wind, and u_j' the deviation of the vertical wind from the vertical mean wind $\overline{u_j}$, respectively. The three dimensional wind components have to be rotated into a perpendicular coordinate system pointing towards the main wind direction[3] before calculating the variances and the resulting friction velocity.

The theoretical discussion on the interactions between solid earth and atmosphere gives attention to forces working at the interface. Subdued to the

[3] described by Wilczak et al. (2001) as planar fit method

equilibrium of forces the sum of applied forces between atmosphere F_{atm} and ground F_{ground} interact:

$$F_{atm} - F_{ground} = 0. \tag{2.7}$$

Reducing the forces from the atmosphere to dynamic forces applied to the ground, analogue to Equations 2.3 and 2.4 the sum of friction F_f and pressure forces F_p remain:

$$F_{ground} = F_f + F_p. \tag{2.8}$$

Assuming a plane surface (Figure 2.5), the force resulting from the horizontal wind flow at the interface, are reduced to the friction force, depending on shear stress applied to the total area A:

$$F_{ground} = F_f = \tau \cdot A. \tag{2.9}$$

Deriving the friction from Equations 2.5 and 2.6 provides a very stationary reproduction of the factual situation at the location of the measurements. When estimating the friction force according to Equation 2.9 for a broader area, the approach becomes inaccurate, disregarding obstacles, topography or change of friction parameters of the surface that may lead to dynamic pressure changes. Then, the dynamic pressure in Equation 2.2, term (III) can not be neglected any more leading to the description in Equation 2.8.

Limitations of Model Assumptions of Shear Described by Friction Velocity

The consideration of u_i' and u_j', respectively the friction velocity u_* (Equations 2.5 and 2.6) as parameters for the shear stress of the mean wind field is valid for the assumption of a parallel, horizontal mean air flow. It is predominantly satisfied for measurements above homogeneous, plain terrain. Thereby, the inertial system needs to be rotated into the mean wind direction, parallel to the surface (Wilczak et al., 2001).

Elevated terrain causes topographical forcing of air motion, such as an uplift for upslope winds, causing significant mean vertical wind speed and systematically altered u'_j. The calculations of the friction velocity and the shear according to Equations 2.5 and 2.6 are then affected systematically. The calculated shear stress does not solely represent the friction above the surface then, but is also affected by changes of turbulence due to different stratification of the boundary layer. Carlson and Foster (1986) conduct numerical simulations of wind flows perpendicular to a valley. Their Carlson model predicts reduced surface stresses for downslope winds and enhanced surface stresses for upslope winds compared to wind across plain terrain. Smith (1985) simulate transitional flow downslope of mountains, finding an acceleration of the surface winds and attending reduction of the wind shear. A brief discussion on the influence of topography on wind flow and the boundary layer can be found in Stull (1988). Wind shear and shear stress thus do not necessarily solely depend on the frictional characteristics of the surface, also they are inflicted by dynamic pressure changes due to topographic forcing.

Using common mechanical approaches, it is possible to estimate the forces from the atmosphere applying to the solid surface. As convenient approach of a plane surface the friction velocity is considered in this thesis. The forces applied at the surface equal the forces that cause signals or noise in seismological records such as ground motion velocity. For meteorological matters, it is possible to get an estimation of the effective forces from stationary, turbulent measurements providing the friction velocity.
The corresponding approach in seismology deriving the contact forces from seismological measurements is challenging and probably not possible using present model approaches (cf. Section 2.1). This study, instead of the theoretical idea, develops a more practical approach and finds a scaling dependency of the seismological noise level, respectively wind-related signals, with the mean horizontal wind speed as a most simple approximation.

The derivations described in Equations 2.5 to 2.9 can be deduced from common fluid mechanical approaches (i.e. Brandt and Dahmen, 2005) and literature on boundary layer meteorology (i.e. Stull, 1988) in greater detail. In the case of a gradient of the ground or more complex terrain, the calculations of the perpendicularly defined, turbulent deviations do not represent the factual turbulent deviations perpendicular to the ground (Lee et al., 2004). The horizontal wind components, defined as parallel to the ground, may be influenced by changed stratification, which is further discussed in Chapter 7.2.1.

2.3 Seismic Array Processing

A Seismic array, consisting of multiple, geometrically linked seismological stations, can be used to increase the signal-to-noise ratio of the single records by stacking signals and treating the array as one recording station. Further, the knowledge on the locations of the stations allows to measure properties such as the epicentral directions of propagating waves in the earth. The locations of the seismological stations, deployed in the scope of this thesis are chosen as to fulfil requirements of array processing (cf. Chapter 4). This section briefly outlines the geometrical considerations of array seismology, following more elaborate derivations by Rost and Thomas (2002); Schweitzer et al. (2012).
For distances from wave source to receiving station, that are significantly larger than about 10 wave lengths, the seismic wave can be approximated as a planar wave front crossing the array (Almendros et al., 1999). Thereby, the vectorial velocity of the wave can be described by two characteristics: the backazimuth ϕ, which is the horizontal angle of wave front approach, and the horizontal slowness u, the inverse of the horizontal projection of the wave front velocity. Figure 2.6 schematically illustrates a wave front crossing seismological stations of a seismic array (a). The wave front reaches each of the stations with a specific time delay t_i with respect to the approach at the array centre t_0. Thereby, the wave front covers the specific distance d_i (b).

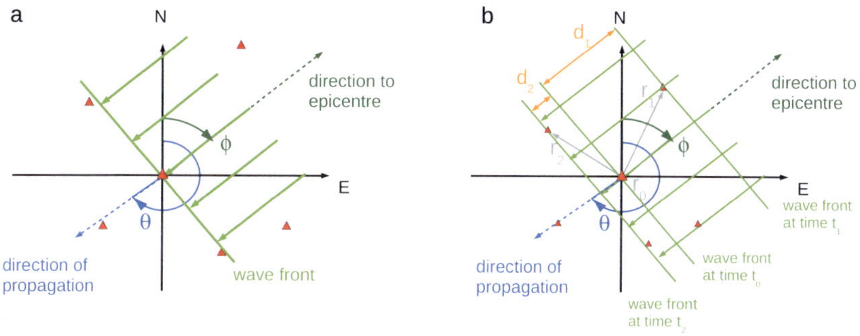

Figure 2.6: Illustration of a plane wave propagating across a seismic array as horizontal projection, relative to north (N) and east (E) direction. **a:** Definition of wave characteristics approaching the array. The wave front is perpendicular to the direction of the propagation. Red triangles mark seismological station locations r_i. The array centre is defined for a specific seismological station set to r_0. The direction of propagation is described by the backazimuth ϕ between north and the direction to the epicentre, or the azimuth θ between north and the direction of propagation. **b:** The wave front at specific arrival times t_i at the station locations r_i. d_i is the relative, horizontal distance covered by the wave front.

Based on the time delays t_i and the relative station locations r_i of the wave at each location several methods to measure the slowness and the backazimuth from array records have been developed. The frequency-wavenumber analysis (fk-analysis), used in Chapters 4.1.5 and 6.4, is one technique to estimate the wave characteristics from array records. The fk-analysis simultaneously calculates the distribution of power among different apparent wave velocities v_{app} and directions of approach (Capon, 1973; Harjes and Henger, 1973; Aki and Richards, 1980). Thereby, a grid search for best fitting time delays, respectively backazimuths and slowness values, is conducted. The seismic records at all receivers are stacked and time-shifted. Computing in the spectral domain, produces the highest energies for time delays t_i that best fit the backazimuth and the slowness of the incoming wave. The station i with the location r_i records the time series:

$$x_i(t) = f(t - r_i \cdot u) + n_i(t). \tag{2.10}$$

Thereby, f is the signal related to the wave front and n_i the noise signal at each individual station. u is the horizontal slowness vector

$$u = \frac{1}{v_{app}} \cdot (\cos\phi, \sin\phi), \tag{2.11}$$

as apparent from geometrical considerations in Figure 2.6. The maximum amplitude of the sum of all seismic recordings is reached if all signals are in phase. Therefore, the time shifts between the signal arrivals have to disappear:

$$t_i = u \cdot r_i = 0. \tag{2.12}$$

A trace with removed time shift can be calculated by

$$\tilde{x}_i(t) = x_i(t + r_i \cdot u) = f(t) + n_i(t + r_i \cdot u). \tag{2.13}$$

The beam trace as the sum of all shifted recordings results from

$$b(t) = \frac{1}{N} \sum_{i=1}^{N} \tilde{x}_i(t) \tag{2.14}$$

$$= f(t) + \sum_{i=1}^{N} n_i(t + r_i \cdot u). \tag{2.15}$$

Using Parseval's theorem, the total energy of the finite signal can be derived from the beam $b(t)$. By integrating the squared summed amplitudes over time the total energy

$$E(k - k_0) = \int_{-\infty}^{\infty} b^2(t) \mathrm{d}t \tag{2.16}$$

$$= \frac{1}{2\pi} \int_{-\infty}^{\infty} |S(\omega)|^2 \underbrace{| \frac{1}{N} \sum_{n=1}^{N} e^{2\pi j \cdot (k - k_0) \cdot r_i} |^2}_{ATF} \mathrm{d}\omega \tag{2.17}$$

depends on the Fourier transform $S(\omega)$, depending on the angular frequency ω of the signal s(t). k thereby is the wave number vector

$$k = \omega \cdot u \tag{2.18}$$
$$= \frac{\omega}{v_{app}} \cdot (\cos\phi, \sin\phi). \tag{2.19}$$

The term ATF in equation 2.17 includes all array characteristics, independent of the incoming wave. It can be abstracted as:

$$|A(k-k_0)|^2 = |\frac{1}{N}\sum_{i=1}^{N} e^{(s\pi j(k-k_0)\cdot r_i)}|. \tag{2.20}$$

It is the array transfer function (ATF) that controls the total energy recorded at the seismic array. The ATF depends on the array aperture and the inter station distances between the individual stations. The array specific ATF describes the sensitivity of the array to waves with specific slowness vectors and backazimuths. It is therefore essential to consider the ATF in order to conduct fk-analysis and to calculate the direction of an incoming wave.
Horizontal slowness (Equation 2.18) and backazimuth can be associated to the absolute value of the wave number vector:

$$|k| = \sqrt{k_x^2 + k_y^2} = \frac{\omega}{v_{app}}, \tag{2.21}$$

$$\phi = \tan^{-1}\left(\frac{k_x}{k_y}\right). \tag{2.22}$$

The fk-diagram displays the power spectral density in polar coordinates with respect to the backazimuth on the azimuthal axis and the wave number on the radial axis. It can be used to illustrate the ATF of an array or the power spectral density of an array under the influence of an incoming wave. Unambiguous energy maxima in the fk-diagram thereby indicate the backazimuth and the slowness of the incoming wave.

The performance of the ATF depends on the locations of the seismological stations of the array. Thereby, inter station distances, maximum station distances, and the azimuthal arrangement of the stations determine the distribution of power along the wave number grid. The station distances dictate for which frequencies the seismic array is sensitive and appropriate without ambiguity due to aliasing effects (cf. Chapter 4.1.5). The array setting therefore has to be chosen according to the purpose. A profound discussion of several examples of feasible array designs can be found in Schweitzer et al. (2012).

3 The Dead Sea Region and the DESERVE project

The survey area of this study is a particularly selected part in the Dead Sea region. To illuminate the special background of the Dead Sea region in terms of geology and climatology, in fact in a state of interdependence, the following sections give a brief overview about the region and the DESERVE project.

3.1 The Dead Sea Region

The Dead Sea is a hypersaline lake situated at the boarder triangle of Isreal, Jordan and Palestine (Figure 3.1). It lies between the Sea of Galilee in the north and the Red Sea in the south within the depression of the Jordan Rift Valley. For a length of about 50 km along the rift, a width of about 15 km, its shores mark the Earth's lowest land elevations. The lake is fed by the Jordan River in the north and several wadis from surrounding mountains, but there is no discharge into the oceans, hence the high salinity of approximately $277\,\mathrm{g\,kg^{-1}}$ (as of state 2005, Lensky et al., 2005). For decades, the Dead Sea has been shrinking, the former lake divided into a northern basin, the present Dead Sea, and a southern basin, by now separated evaporation ponds, artificially fed with Dead Sea water for industrial purposes.

3.1.1 Geology

The climate and the hydrological cycle in the Dead Sea region are predominantly determined by the geology of the area. The Jordan Rift valley is a segment of the Greater Rift valley, a trench of 6,000 km length extending from

Lebanon to South Eastern Africa. Between Jordan and Israel, the Dead Sea valley forms a deep depression as part of the Jordan Rift valley, evolved from the Dead Sea Transform (DST) fault system. It separates the Sinai Plate with the Israel mountains in the west from the Amman plateau on the Arabian Plate in the east. Evolved from north and eastward movement of the Arabian plate, the deep, steep valley of the Dead Sea formed. On a small scale of 30 km perpendicular to the fault, the valley experiences height differences of more than 1500 m in both directions, the elevations ranging from more than 1000 m in the mountains down to the water level of the Dead Sea, at present state about -430 m with respect to sea level.

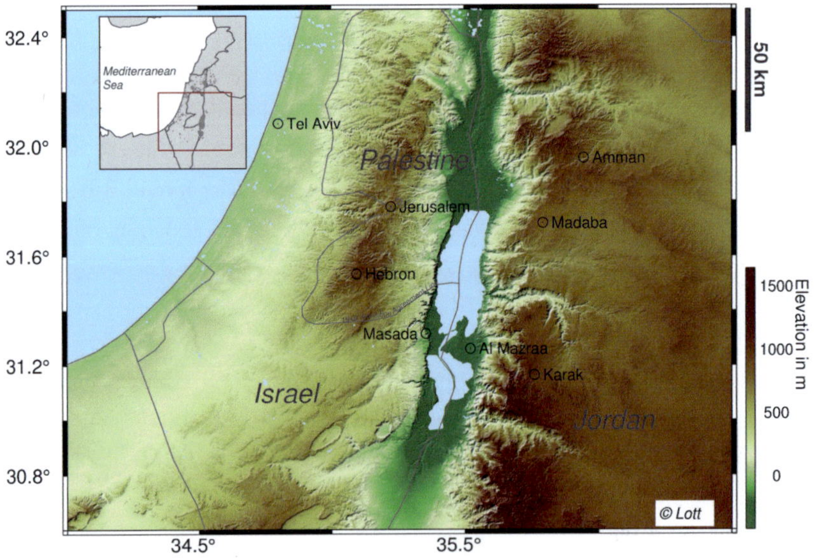

Figure 3.1: Topographical map of the Dead Sea region. The Dead Sea in blue colour in the centre. The separate, southern parts of the Dead Sea are shallow evaporation ponds on higher elevation, linked to the main water body by canals.

3.1.2 Synoptic Situation in the Eastern Mediterranean

The weather of the Eastern Mediterranean is subject to several synoptic systems, analysed for genesis and duration, and postulated by Alpert et al. (2004a), determining the general weather situation. The winter and summer months are controlled by different weather systems. Persian Troughs exists in the summer months and dictate the weather, bringing along high pressure areas and very arid air with almost no precipitation. In autumn and winter storms can occur, characterised by high wind speed and occasional precipitation. Thereby, most storms in autumn can be connected to a situation, classified as Red Sea Trough (RST) by Alpert et al. (2004b). It is often associated with strong, dry wind from the continental, easterly dry sector, rarely generating precipitation (Saaroni et al., 1998; Tsvieli and Zangvil, 2007). Storms in winter can often be related to rain-bearing Mediterranean cyclones significantly influencing the weather in the Eastern Mediterranean with winds from west (Alpert et al., 1990).

3.1.3 Climate in the Dead Sea Region

Naming the Dead Sea region after the water body of the Dead Sea accounts for the fact that the environmental situation in the region is closely linked to the Dead Sea. Thereby, the dependence is mutual. While the Dead Sea controls and regulates the local climate, at the same time it highly depends on the climate as well as on anthropogenic developments in the region. The environmental processes in the region have been in a state of a very sensitive equilibrium that has gotten out of balance since the 1960's when the drop of the Dead Sea water level rapidly accelerated. Potash industry at the southern basin of the Dead Sea enforcing rapid evaporation of the Dead Sea water in combination with man-made shortage of the water inflow from the Jordan River has led to an approximate drop of the water level of 1 m per year with collateral, fatal implications for the environment (Krumgalz et al., 2000; Yechieli et al., 1998).

The climate in the Dead Sea region can be described as a transition zone between semi-arid in the mountains and arid in the valley (Kafle and Bruins, 2009). In summer, the desert environment undergoes strong diurnal temperature gradients. The different levels of exposure of mountains, valley slopes and valley bottom to solar insolation lead to strong local temperature gradients. Especially in summer, distinct, local winds in and around the valley evolve, interacting and often decoupling the valley flow from the synoptic wind field above (Alpert et al., 1982).

Figure 3.2 shows a conceptual model for the circulation of wind systems in summer perpendicular to the long axis of the Dead Sea valley, based on Herzog (2015). Synoptic wind, dominating from west, streams above the Dead Sea valley and depending on the time of day, interacts with local wind systems inside the valley. At elevations between mountain tops and valley bottom, an inversion may evolve, leading to the decoupling of the wind regime at the bottom from the synoptic scale flow above. On the one hand, it implies, that the wind regime inside the valley depends on the elevation, involving changing wind speed and wind direction. On the other hand, the decoupled wind system itself is small-scale, possibly rotating, and thereby changing wind direction within short distance, bounded by the dimensions of the valley.

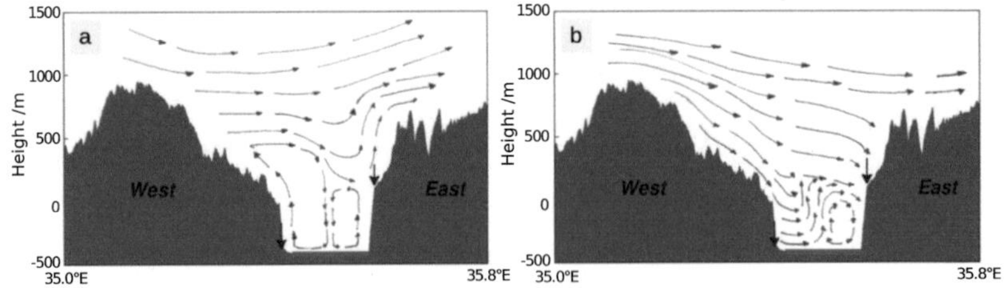

Figure 3.2: Conceptional model of wind systems in the Dead Sea valley in summer, illustrated in cross sections from west to east. Red arrows mark sites of KIT meteorological stations MET1 and EBS3. Topography is raised. **a:** day time. **b:** night time. Wind arrows taken from: Herzog (2015).

As a result, the wind can strongly vary in horizontal and vertical direction, for example at facing valley slopes in the east and west, or for different elevations above the valley bottom. The topography of the valley confines the processes to very complex, local occurrences that can not be sufficiently resolved by current weather models, such as the Global Forecasting System (GFS) with a regional resolution of 0.5°.

3.2 The DESERVE Project

Since 2012, the Dead Sea region is subject to the Helmholtz Virtual Institute Dead Sea Research Venue (DESERVE). In a multidisciplinary approach the project addresses the challenges of the rapidly changing environment at the Dead Sea using geophysical, meteorological and hydrological expertise, summarized in Kottmeier et al. (2016). Participating institutes from the Helmholtz-Association are the German Research Centre for Geosciences Potsdam (GFZ), the Helmholtz Centre for Environmental Research (UFZ), and the Karlsruhe Institute of Technology (KIT), each involved with associated partner institutes in Israel, Jordan and Palestine.
In the scope of the DESERVE project, an extensive, multidisciplinary monitoring network is established, going along with multiple temporary field campaigns, conducted in and around the Dead Sea valley. Figure 3.3 shows an excerpt of the station network, including meteorological stations deployed by the Institute of Meteorology and Climate Research (IMK) and one seismological permanent station, deployed as part of the GEOFOrschungsNetz (GEOFON) network by the GFZ and the associated partner, the Ministry of Energy and Mineral Resources (MEMR) in Jordan.

The seismic array close to the north eastern shore of the Dead Sea, marked as red circle, provides the foundation for this study. Installed as a cooperation of the Geophysical Institute (GPI) at the KIT, the IMK, and local partners from the MEMR with kind support from GFZ, the measurements provide the basis to investigate the influence of wind on seismological records in this study. The run time of the temporary seismological stations from February, 2014 to February, 2015 coincides with the deployments of the meteorological stations EBS3 and MET1, the latter installed intentionally to accompany the seismological array. GHAJ is the closest permanent seismological station in service during the deployment of the seismological array and used as reference in this study.

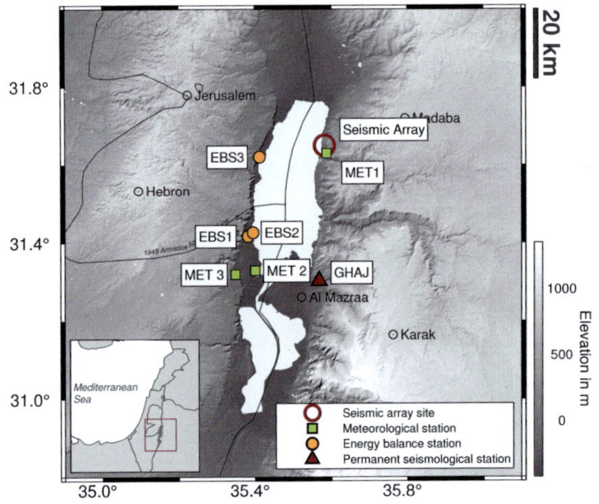

Figure 3.3: Map of the Dead Sea area illustrating all meteorological stations installed by IMK, a permanent seismological station installed by GFZ, and the temporary seismic array installed for this study.

4 Experimental Setup, Instrumentation and Data Quality

This work explores the interaction of wind and ground movement based on a target oriented geophysical experiment. Therefore, the base for all further research is real field data, coming along with common advantages and disadvantages of experimental raw data. It is important to consider the circumstances under which data was maintained. As a foundation for following analyses (Chapters 5 to 7) this chapter describes the equipment and the obtained data. The experiment was designed, conducted, and analysed by the author of this work.

Figure 4.1 gives an overview of the locations of all relevant measurements for this work that are discussed in detail in this chapter. The sites of all measurements are located close to the Dead Sea and well within the Dead Sea valley, however on different elevations. With the exception of the meteorological eddy-covariance station EBS3, which is located directly on the western shore in Palestine, all sites are east of the Dead Sea in Jordan. The main site of this work is in the north-eastern part of the Dead Sea valley between the city of Madaba and the Dead Sea shore where meteorological Station MET1 and the temporary seismological stations were installed.
The following sections include information on the site and data of the seismological array built for this project (Chapter 4.1), meteorological data obtained by IMK (Chapter 4.2), and seismological data from recording station GHAJ from the GEOFON network in Ghour Hadithe (Chapter 4.3).

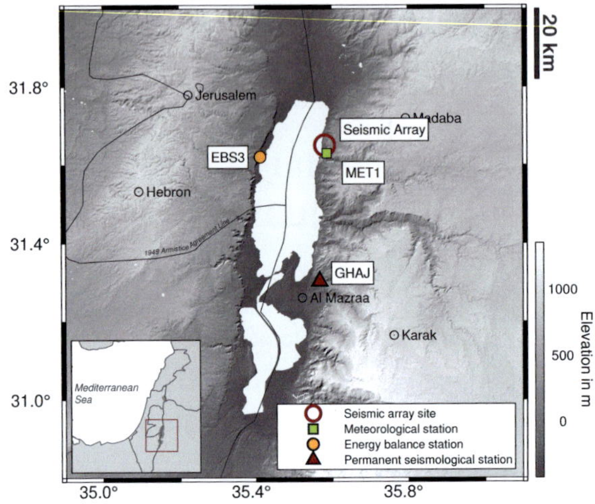

Figure 4.1: Map of the Dead Sea illustrating locations of measurements conducted for or used in this work. Topography is illustrated by color and shaded relief. Source of digital elevation model: SRTM data from Consortium for Spatial Information (CGIAR-CSI) (Jarvis et al., 2008)

4.1 Seismic Array

To estimate the quality and the resolution properties of seismological data it is important to consider the site of each individual station, affecting the data quality, and the instrumentation that defines characteristics of the data, such as the instrument responses (Figures 4.6 and 4.7 in Chapter 4.1.3). The selection of locations for the temporary seismological stations follows a specific design to enable the application of array seismology methods. At the same time the individual station locations have to conform requirements to achieve good data quality, particularly a good signal-to-noise ratio, and to capture desired signals.

We installed 15 temporary seismological stations, consecutively named from DES01 to DES15 for an intended run-time of one year at the slope of the Dead Sea valley between the Dead Sea shore and a plateau of higher mountains in the direction of the city of Madaba. A brief summary of the seismological measurements and the obtained data in the scope of this study is given in Lott et al. (2016). The recording stations are arranged as one seismic array with spatial extent of 2.7 km and can be processed as multiple independent stations

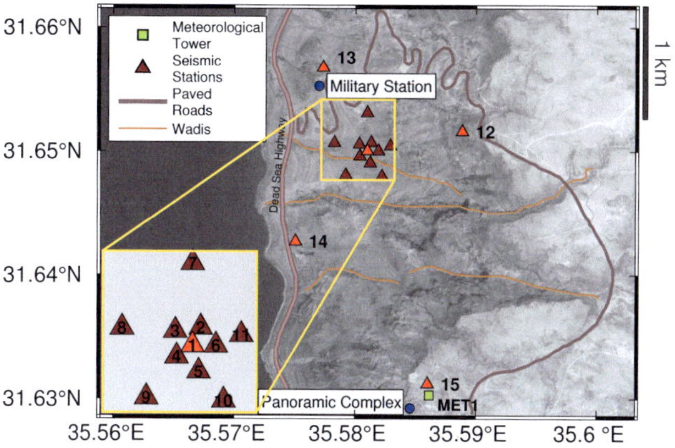

Figure 4.2: Map covering the area of the seismological array. All temporary seismological stations are marked as red triangles and numbered according to the station names DES01 to DES15. Blue circles indicate buildings. Source of satellite image data: Bilder@CNS2015 / Astrium, DigitalGlobe

or treated as seismic array. Figure 4.2 shows the locations of all seismological stations marked as red triangles. As the satellite image indicates the stations are located in steep and highly ragged terrain (see also Figure 4.15). All stations are arranged around the central Station DES01 (mark 1 in the map). Basically, the station's locations form three circles around DES01 consisting of each four or five stations:

- Circle 1: 5 Stations DES02 to DES06
- Circle 2: 5 Stations DES07 to DES11
- Circle 3: 4 Stations DES12 to DES15

Though being arranged on circles the station locations are selected as to be evenly distributed in the area. Table 4.1 gives an overview on the seismological array sites as table. In addition to the station locations also station instrument parameters, such as the sensor type are listed. Two stations not

Table 4.1: Locations, instrumentation and run times of seismological stations. Note the mostly negative station altitudes due to the elevation of the Dead Sea valley below sea level. Further instrument details are explained in Chapter 4.1.2.
* MP: Mark Products, TC: Trillium Compact

Station	Location	Alt. /m	Sensor, ID	Logger, ID	Runtime
DES01	31.65002N,35.58100E	-292	TC* 015	CUBE 710	2014/02/22-2014/11/30
					2014/12/09-2015/02/12
DES02	31.65068N,35.58132E	-189	MP* 1357	CUBE 645	2014/02/25-2015/02/12
DES03	31.65055N,35.58033E	-209	MP* 1824	CUBE 641	2014/02/25-2015/02/11
DES04	31.64960N,35.58038E	-232	MP* 1885	CUBE 642	2014/02/27-2015/02/11
DES05	31.64904N,35.58124E	-276	MP* 1871	CUBE 643	2014/02/27-2015/02/11
DES06	31.65001N,35.58192E	-263	MP* 1354A	CUBE 644	2014/02/27-2015/01/05
DES07	31.65307N,35.58100E	-190	MP* 1879	CUBE 640	2014/02/22-2015/02/12
DES08	31.65066N,35.57826E	-280	MP* 1823	CUBE 646	2014/02/24-2014/10/18
					2014/10/31-2015/02/11
DES09	31.64807N,35.57920E	-257	MP* 1355A	CUBE 647	2014/02/24-2014/06/20
					2014/08/08-2015/02/11
DES10	31.64796N,35.58222E	-195	MP* 2831	CUBE 648	2014/02/24-2015/01/05
DES11	31.65041N,35.58291E	-162	MP* 1898	CUBE 649	2014/02/24-2015/01/30
DES12	31.65158N,35.58886E	90	TC* 013	CUBE 711	2014/02/26-2015/02/12
DES13	31.65673N,35.57734E	-248	TC* 016	CUBE 712	2014/02/26-2015/02/11
DES14	31.64270N,35.57505E	-374	TC* 014	CUBE 713	2014/02/26-2015/01/05
DES15	31.63120N,35.58603E	123	TC* 007	CUBE 650	2014/02/23-2015/02/17
DES20	31.63028N,35.58616E	123	TC* 016	CUBE 712	2015/02/11-2015/02/17
DES21	31.63028N,35.58616E	123	TC* 014	CUBE 713	2015/02/11-2015/02/17

marked in the map in Figure 4.2 are listed as well. Stations DES20 and DES21 were installed at the end of the field experiment for the period of one week for reference measurements.

4.1.1 Experiment Design and Station Sites

The seismological and meteorological measurements in Jordan started in February, 2014. The seismological stations were unmounted one year later, but the meteorological station MET1 at the site of Panoramic Complex remained in operation (Chapter 4.2). The installations and first maintenance cycle were conducted within four weeks in cooperation with the GPI at KIT and the local partner MEMR, former National Resources Authority in Jordan (NRA), who provided expertise and manpower during installation and operation of the measurements.

The seismological stations were installed between 22nd February and 27th February, 2014 and dismounted about one year later between 11th and 17th February, 2015. About one week after installation, all stations were maintained and checked for potential malfunctions or data recording problems by the author of this work. After the first test run, that confirmed the functionality of the stations, the local partners in Jordan were in charge of running and maintaining all seismic stations regularly till the end of the experiment.

Setting of Seismological Stations and Description of the Sites

At some sites where the local conditions of the ground allowed to excavate a hole, the seismometers were deployed into the ground to improve the signal to noise ratio and to shield the sensors against direct impact of wind. As to increase the coupling of the sensors to the ground all seismometers were deployed on flagstones inside the holes. The seismometers were levelled and oriented towards the magnetic North (Chapter 4.1.3). Two persons double-checked the orientation and the water level at the instruments during installation to avoid inadvertence related mistakes. All seismometers were covered by plastic tarpaulins and domes that were re-covered with soil and stones to isolate the sensors against thermal fluctuations, protect it against rain and wind, and to hide the installation visually. Panel a in Figure 4.3 illustrates the schematic installation of a seismological station that was deployed below the surface into the ground, such as station DES01, for example. Figure 4.4 summarizes and documents the installations of all seismometers during or after deployment.

The data loggers linked to the seismometers and the batteries were installed close by. As seismometers need an undisturbed and firm connection to the ground and as little influence from its surroundings as possible, the data loggers need a direct connection to the GPS satellites. Hence, the data loggers

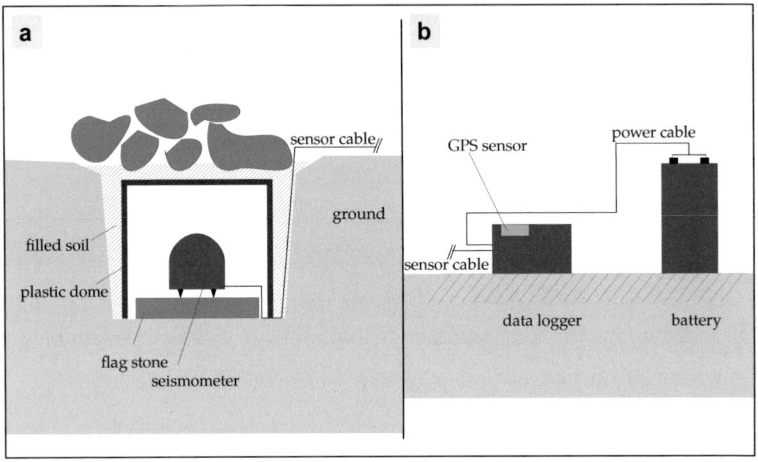

Figure 4.3: Schematic design of seismological station. **a:** Seismometer, deployed on a flagstone in the ground, shielded by isolation material and a plastic bucket, and covered by soil and stones. Note, that the seismometer is decoupled from the bucket. **b:** Data logger and battery, nearby. The data logger needs to be placed in open space to be able to connect to GPS.

were positioned separately above the ground, only connected to the sensor by a cable, as illustrated in Figure 4.3b. Moreover, during exchange of the batteries and the data collection, the seismometers remain undisturbed except for the seismic noise caused by human activity in the vicinity.

Except of rare small plants common for steppes and desert areas in the Saharo-Arabian region (cf. Al-Eisawi, 1996) there is no vegetation at the site. Whereas above sea level, especially in winter, occasionally some seasonal plants occur, with decreasing elevation vegetation vanishes except for local spots such as wadis where springs may even feed date palms. The lack of vegetation implicates eroded, bare ground in a rugged area and sometimes bedrock. The area of the field experiment is a very steep terrain and by reason of its location at the slope of the Dead Sea valley also rugged by small wadis running through the site and separating the recording stations.

Table 4.2: Characteristics of seismological stations. For an estimation of the data quality sensor depth, potential noise sources and close obstacles are listed for each station. Most relevant in matters of anthropogenic noise would be the Dead Sea highway close to Station DES14.
* Approximate designation.

Station	Alt. /m	Sensor depth* /cm	Pot. noise sources	Characteristics/Obstacles
DES01	-292	40	-	Plain site
DES02	-189	35	-	Between stones
DES03	-209	40	-	-
DES04	-232	20	-	Between rocks, Close to wadi in the south
DES05	-276	40	-	Plain site
DES06	-263	0	-	-
DES07	-190	40	Minor road	Big rock to the west
DES08	-280	40	Minor road	Plain site
DES09	-257	20	Wadi	Station embedded between rocks
DES10	-195	20	Wadi	Station embedded between rocks
DES11	-162	0	-	Close to rock
DES12	90	40	Minor road	Steep slope of terrain
DES13	-248	40	Military station, Minor road	Close rock west of station
DES14	-374	40	Dead Sea highway	Close rock west of station
DES15	123	0	Met. station, Panoramic Complex	Plain site
DES20	123	30	Met. station, Panoramic Complex	Sensor not on flagstone
DES21	123	30	Met. station, Panoramic Complex	Sensor not on flagstone

The seismological instruments were deployed in an area of 2.7 km diameter at maximum at altitudes ranging from 123 m above sea level to −374 m below sea level. The conditions of the ground didn't allow to deploy all seismometers below surface at predetermined sites, therefore the site conditions at various stations may differ and cause individual site effects, such as different noise levels in the recordings. Table 4.2 (cf. photographs in Figure 4.4) subsumes relevant conditions, such as the depth of the seismometers and potential noise sources at all stations. DES01, 03, 05, 07, 08, 12, 13, and 14 are recording stations where the soil conditions allowed to dig the seismometer including its plastic dome with full coverage, at a depth of at least 40 cm below surface. Seismometers at other stations were only partly buried or deployed directly on solid rock at the surface, such as seismometers at DES06, 11, and 15. At stations that are not fully buried in the ground one has to expect a higher noise level (Trnkoczy et al., 2012). To counteract, in addition all seismometers of the setting were covered by domes and deployed at sites that were favourably wind-shielded by rocks around the seismometer installations. Domes that were not completely beneath the surface were covered and surrounded by stones and small rocks to avoid direct exposure to wind and sun.

All station sites except of Station DES15, and close by DES20 and DES21 respectively, are located in an unguarded, open area kept aloof, but still close to the highly frequented Dead Sea Highway (Figure 4.2) running along the east coast of the Dead Sea, and close to a winding road from the Dead Sea up to the city of Madaba in the east. Five km north of the measurements at the Dead Sea Highway there is a complex of several hotels, frequented by tourists and local visitors most excessively in spring and autumn. As strong sources of noise directly in the area of measurements one can expect anthropogenic influence as well, from the a military station close to Station DES13 and the Panoramic Complex where Station DES15 (and later DES20 and DES21) were installed. The military station consists of only one small building with only a few persons in charge at the same time. However, we can expect them to have a generator as electric power source on-site causing noise in the seismological recordings. Compared to the military station the Panoramic Complex including visitor center, museum, restaurant, viewing platform, parking, and a solar power plant, takes a much bigger area and attracts more people and traffic, therefore probably influencing the seismological records in a more severe way. Like at the roads, we expect the noise level there to vary by time of day, day of week, and for special occasions, correlating with working hours, work days, and holidays or festivities (Section 4.1.4).

Table 4.3: Description of potential noise sources close to Station DES15 at Panoramic Complex. For each obstacle the approximate distance to Station DES15 is listed. Local time corresponds to UTC+2 h in winter, local daylight saving time to UTC+3 h in summer.

Obstacle	Distance /m	Signal source	Occasion
Museum, Visitor centre	190	anthropogenic, electric generators	during day times open 9:00 to 17:00 local time
Restaurant	200	anthropogenic, generators	9:00 to 22:00 local time
Parking, Street lamps	160	traffic during strong winds	day times
Solar power plant, Generators	230	electric generators	all times of day
Road	350	traffic	mostly day times
Farm building	180	anthropogenic, pumps	-
Farm land	30	pumps	-
MET1	100	vibration	during strong winds

Station DES15 is significantly exposed to anthropogenic noise originating from the Panoramic Complex. However, it is one of the most important stations of the array considering that it is the closest one to the meteorological station MET1, and at the same elevation. At Station DES15 one can expect the same local wind conditions as at MET1. In contrast, at all other stations the local wind speed cannot be assumed identical in both amount and direction due to severe changes in elevation and the rugged terrain channelling the wind near to the surface (Chapter 3.1.3, Figure 3.2). Therefore, it is advisable to consider potential noise sources at DES15 in detail.

The small distance of about 100 m of DES15 to MET1 makes the wind conditions comparable, even on local scale. However, it brings along the disadvantage that due to the vicinity the seismometer at DES15 might record the vibration of the meteorological tower itself. As to give an estimation of the type, occurrence and intensity of noise sources Table 4.3 lists potential noise sources close to DES15. The distance of the source can lead to a rough estimation on the intensity of the influence at the station, whereas the kind of

source can lead to signal properties such as the frequency band of the noise. When analysing data from DES15 knowledge about noise sources can help to identify signals. The vicinity of the facilities belonging to the Panoramic Complex indicates anthropogenic noise at DES15 especially during opening hours, or rather between approximately half an hour before opening and half an hour after closing, from 09h to 17h local time.
For the interpretation of seismological signals it is important to know about potential noise at the recording stations, especially when investigating broad frequency bands that are affected by multiple noise sources, amongst others the signals of interest. Chapter 6 will show that the same signal sources can cause diverse signals recorded at different stations. One reason are different distances and ray paths from source to receiver, where the signal encounters specific modulation and attenuation due to different specific properties of the ground. Another important reason, however, is that the recorded signals are strongly modulated by unwanted site effects, leading to interferences with local signals from the site of the sensor. Therefore it is essential to consider the characteristics of the station's sites, too when analysing seismological signals.

Station Maintenance and Deficit of Data

The experiment was designed to acquire ground motion velocity with continuous recording throughout one year. However, the limitation of internal storage at the data loggers and the confined capacity of the batteries required station maintenance at all stations by a regular schedule approximately every four weeks to eight weeks. During each procedure the data acquisition was interrupted leaving gaps and noise from the disrupting procedure in the otherwise continuous records.

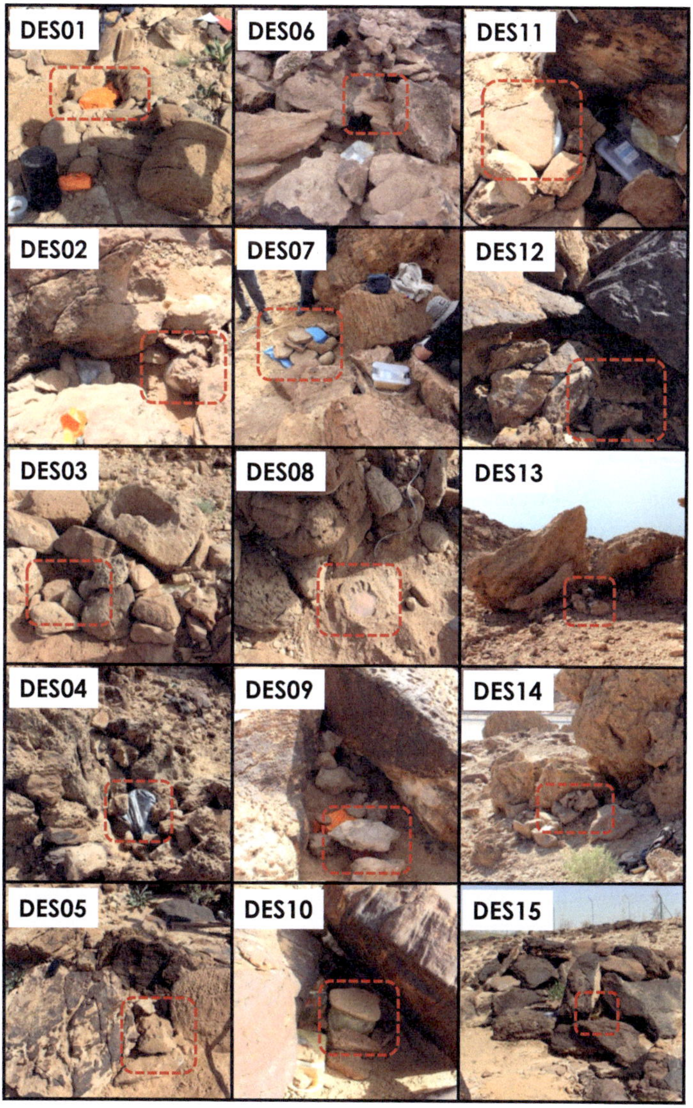

Figure 4.4: Photographs of all DES stations during installations. Red frames mark positions of seismometers. Pictures of Stations DES01, DES04, DES07, DES08, DES09, and DES10 were taken before seismometers were covered entirely.

After the installations in February 2011 and the first station maintenance local workers from MEMR pursued the maintenance procedure by a given schedule till the end of the experiment. During the recording period of one year the data have a high retrieval rate of 96% with almost no losses. Figure 4.5 illustrates the availability of seismological records acquired in this experiment for each station. Stations were deployed and unmounted on several consecutive days, therefore data acquisition periods vary for each station marginally (station run times in Table 4.1). Grey squares mark instances where station maintenance was conducted, causing gaps of approximately 30 minutes. The data also reveal three incidents, where the back up routine during maintenance obviously didn't work correctly, leaving long gaps of nine days to more than one month in data of Stations DES01, DES08 and DES09. However, along with the consecutive maintenances the data acquisition continued successfully. Three Stations DES06, DES10, and DES14 stopped recording before the end of the experiment. In January, 2015 they were discovered by locals and collected by the police. Apart from the last month of the experiment it is possible to use data from nearly all stations.

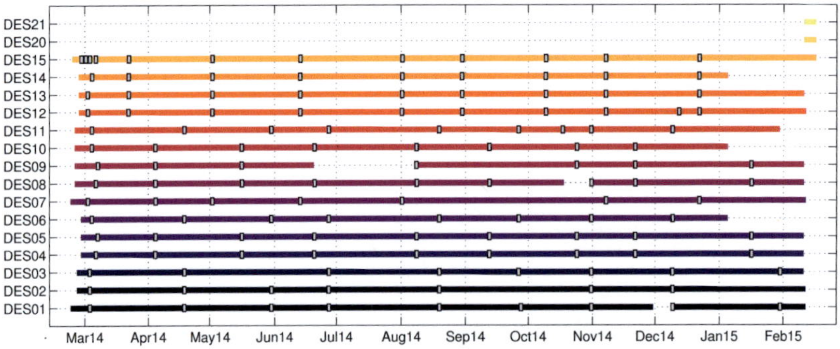

Figure 4.5: Availability of seismological data during the whole measurement period. Grey squares mark small data gaps caused by station maintenances.

4.1.2 Equipment

There are several requirements demanded on seismological stations deployed in the open, unguarded field for several months without supervision. The instrumentation of the seismological stations including sensor, data recorder, and power source, has to be selected accordingly.
In the open field no external power source is available and solar panels to power the instruments are inappropriate due to their high visibility in the unguarded sites. Therefore, instruments used in this experiment were chosen as to have only low power consumption to be fed by batteries over long periods. All seismological sensors and data loggers including data and power cables, used in this experiment, were loaned devices provided by the Geophysical Instrument Pool (GIPP) at GFZ. We installed two types of seismological stations consisting of different sensor types. Five stations were broad band stations equipped with Trillium Compact 120 s seismometers, ten stations used 1 s Mark L4-3D seismometers. Calcium car batteries with a capacity of 35 Ah served as power source for all seismic stations. The capacity of the batteries was limited to approximately six to eight weeks of run time with our settings before exchange and recharge were required.

Seismometers

The experimental setting was designed to built on two different types of seismological stations for the purpose of array processing. The central Station DES01 and the 4 outmost Stations DES12, DES13, DES14, and DES15 were equipped with broad band instruments, the remaining 10 stations were equipped with short period seismometers. Table 4.1 lists the instrumentation including seismometer type of each seismological station. The specific characteristics of both instrument types are subject to the following paragraphs and further discussed in Section 4.1.3 concerning their data properties.

Table 4.4: Sensor and logger properties for recording stations with Nanometrics Trillium Compact and Mark Products sensors. Note: sensitivity, normalization factor, and poles and zeros in this table are average values for Mark Products instruments. The sensitivity of 750 Vsm^{-1} of the Trillium Compact instrument is divided by the factor 10 due to a divisor in the break out box of the instrument.

Sensor	Logger gain	A-D conversion $/\mu Vcount^{-1}$	Sensitivity $/Vsm^{-1}$	Norm. factor	Poles	Zeros
Trillium Compact	1	0.244	750/10	8.184e+11	-0.0369 + 0.0371i; -0.0369 - 0.0371i; -3.7120e+02 + 0.0000i; -3.7390e+02 + 4.7550e+02i; -3.7390e+02 - 4.7550e+02i; -5.8840e+02 + 1.5080e+03i; -5.8840e+02 - 1.5080e+03i	0; 0; -434.1
Mark Produ	1	0.244	170	1.414	-4.443 + 4.443i; -4.443 - 4.443i	0; 0

Mark L-4C-3D Geophones

Stations DES02 to DES11 were designed as short period stations equipped with Mark L-4C-3D geophones by Mark Products, in the following referred to Mark Products instruments. This geophone type is sensitive to record ground motion velocity in a frequency band between 1 Hz and more than 100 Hz. The instrument characteristics of the Mark Products instrument is well investigated and can be described by its individual instrument response function for each instrument and for each of the three recording components. Table 4.4 lists average poles and zeros for the Mark Products instruments and Trillium Compact instruments. Note, that values listed here for Mark Products instruments are only average values for all instruments. For the data processing in this work, however, individual, more exact values were used (Scriptures A.17 to A.27 in the appendix).

Figure 4.6 shows the average response of the Mark Products instruments for all three components over a logarithmic frequency scale. Above 1 Hz the graph

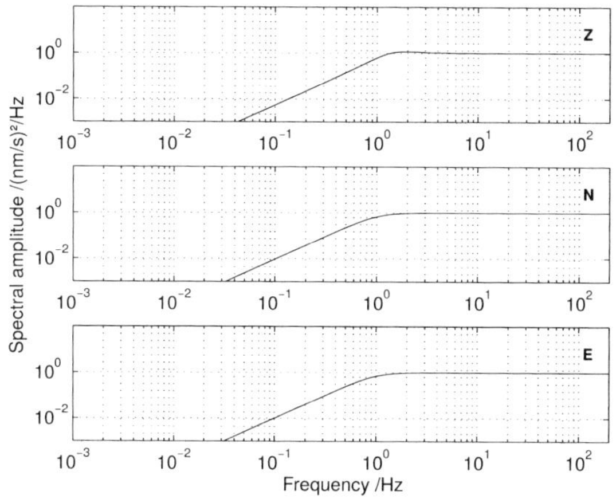

Figure 4.6: Instrument response of Mark Products 1357A at Station DES02. The responses differ only slightly for each of the three recording components and for each of the Mark Products instruments.

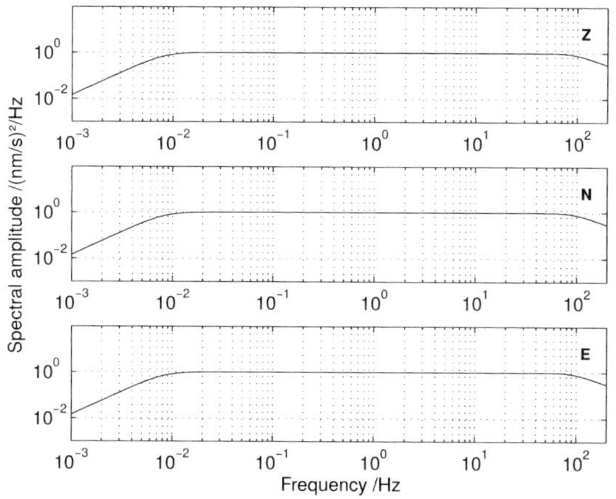

Figure 4.7: Instrument response of Trillium Compact instruments. For all three recording channels and all Trillium Compact instruments the response is the same.

forms a plateau with value 1, where the instrument itself doesn't affect the data recorded. Below, the graph decreases with decreasing frequency, indicating a loss of sensitivity of the sensor in this frequency band. The shape of the instrument response reveals the frequency band in which measurements are reliable and without a significant influence of the instrument itself. For working with frequencies below 1 Hz one should consider removing the influence of the instrument characteristic from the data by applying the instrument response function to the data. This procedure is part of the basic data processing conducted for this work and described in Chapter 5.1.

Trillium Compact 120 s Sensors

Trillium Compact 120 s sensors from Nanometrics, referred to as Trillium Compact instruments, were installed at five seismological stations to allow the array to record broad frequency band recordings. Figure 4.7 shows the instrument response function calculated from the poles and zeros listed for Trillium Compact in Table 4.4. Unlike for the Mark Products instruments, the response is the same for each Trillium instrument and each of its components.

The Trillium Compact instruments are sensitive for periods as long as 120 s, for longer periods the sensitivity of the sensor decays. The graphs form a constant plateau for frequencies higher than 120^{-1}Hz, like the responses for the Mark Products instruments. Above 100 Hz the response function decays. Accordingly, the instrument is sensitive to frequencies between 120^{-1}Hz and 100 Hz without corruption of the original signal. However, it is advisable to remove the instrument response from the data here as well.

CUBE^3 Data Loggers

All stations were equipped with DATA-CUBE^3 data loggers, in the following referred to as CUBE loggers, that recorded the data. CUBE loggers are an in-house development by GFZ, now produced by Omnirecs, with very low power consumption of about 150 mW[1], compared to other devices established. They were originally designed to record seismological data for very short periods of time, such as a few days instead of several months, as it was the case in this experiment. CUBE loggers store data in an internal storage limited to 16 GB in this experiment without option to transmit data while recording.

[1] The Power consumption depends on the chosen GPS settings (in this work: cycled) and the power source (external battery). See further information: http://www.omnirecs.de/dc3.html

The time recording is realized with a built-in quartz oscillator (TCXO) that, according to our experiment set up, had time drifts between −0.005 s per day to 0.04 s per day (Section 4.1.3). For absolute time information and to account for the clock drift the data logger synchronizes the data recording via cycled GPS connection. In our setup (configuration file of CUBE 640 exemplary in Scripture A.16 in the appendix) GPS is connected for a duration of 5 minutes every 30 minutes as a compromise to save power compared to continuous GPS recording. Assuming a continuous clock drift, the procedure provides a time accuracy of ±8.3E-04 s per GPS cycle which by far meets our requirements of ±2.5E-03 s per cycle, respectively 0.5 samples per cycle. As a consequence of the cycled GPS connection, raw data is not evenly sampled and requires the application of a resampling routine in the scope of data processing (Section 4.1.3).
The experiment location in Jordan close to the Israel and Syrian borders was an area of active military presence during our measurements. Due to problems with the GPS signals further described in Section 4.1.3 we have to assume that GPS reception might have been interfered or disrupted, affecting the GPS time signals transmitted to our loggers.

4.1.3 Data Format and Characteristics

CUBE data loggers record seismological data in CUBE raw format that has to be converted into processable data. The GIPP provides a toolbox of programmes including `cube_mseed` by Trond Ryberg (Table A.2 in the appendix) comprising converting and resampling routines. All seismological raw data were converted to MSEED format that can be interpreted by common seismological software and as well be loaded into MATLAB using the open source function `rdmseed.m` (Table A.1 in the appendix).
Data of all stations are available for three recording components, one vertical (Z), and two horizontal components (N and E), each with 200 Hz sampling rate. In raw data, the two horizontal components are not yet corrected for the

geomagnetic declination angle of about 4° in Jordan in 2014 [2]. Therefore, the horizontal components are rotated into more accurate North- and East components according to the Geomagnetic Reference Field as part of the general processing routine described in Chapter 5.1.
The quality of the seismological records highly depends not only on noise sources and site effects but also on the deployment of the station itself. Three main factors congregate: the level of isolation from environmental influence, such as wind or solar radiation, the coupling to solid ground, and the vertical levelling of the seismometer. The first two aspects mainly distinguish temporary stations, such as installed in this experiment, from permanent stations as GFZ station GHAJ, referred to in Section 4.3 or borehole stations. The stations installed as described in this section are well connected to the ground and isolated from environmental influence. However, the data quality won't reach the same level as that of common permanent stations.
During installation, the seismometers were manually levelled using built-in water levels at the seismometers. Not levelling the sensors correctly can cause a drift in the recordings and different sensitivities at North and East recording component of the instrument. This can have significant impact on the energy level recorded at different horizontal components. After one year of recording, the stones the seismometers are deployed on, may have sagged and shifted the seismometers out of their levels. At multiple stations we observed a shift of the water level. At other stations the water bubble leaked making it hard to estimate a drift (Table A.4 in the appendix).
Furthermore, permanent, systematic errors can occur when orienting and positioning the seismometers. For one, inaccuracies or errors can occur when orienting the horizontal components towards magnetic North and East directions. Having double checked the orientation with a compass by two persons during installations, we can exclude significant errors and estimate the inaccuracy of horizontal orientations as below ±3°. The location of the seismometers

[2] Calculation according to International Geomagnetic Reference Field (IGRF). Source: http://www.gfz-potsdam.de/en/section/earths-magnetic-field/services/igrf-declinaion-calculator

was determined using an external GARMIN GPS device. According to the technical report by the NSTB (2014) the global average position domain accuracy of public GPS is ≤ 9 m in horizontal plane and ≤ 15 m on vertical scale. Additionally, the CUBE data loggers record GPS positions along with GPS time every 30 minutes, hence the accuracy is sufficient by far.

Apart from the accuracy of the station locations the timing accuracy is an important aspect to consider when evaluating data quality. As described in Section 4.1.2 the CUBE data loggers have built-in quartz clocks with nearly linear time drift of up to 0.04 s per day. Figure 4.8 reveals the drift of the quartz clocks of all devoted CUBE loggers compared to GPS time[3] over the period of one week, three weeks after installations. Every 30 minutes CUBE data loggers record GPS time. However, the original raw data in CUBE format ignores the GPS time stamps except at the beginning of each day long file and just records with the sampling rate tagged by the quartz oscillator without absolute time information. By ignoring the GPS timing information there would be a time shift of up to 0.04 s each day causing time lapses between each of the day-long files. To counteract and to improve the timing accuracy data has to be tagged and resampled with each GPS time stamp. The software package `cube_mseed` can conduct this process when converting the data from CUBE to MSEED format. However, former releases than the latest release of `cube_mseed` from August, 2015 have problems converting and resampling the data due to wrong and poor GPS reception recorded in the raw data. Still, converting data with the latest release reveals the problem with the GPS signal during the time of the experiment.

Figure 4.9 illustrates the same clock drift as previous Figure 4.8, though for the period of one day only. The clock drift vector $\boldsymbol{\delta t}$ is being calculated by subtracting the resampled, GPS tag-accounted time vector from the uncorrected time vector as given by the quartz oscillator:

$$\boldsymbol{\delta t} = \boldsymbol{t}_{GPS} - \boldsymbol{t}_{clock}. \tag{4.1}$$

[3] Time transfer domain accuracy: 40 ns within 95% confidence interval. Source: NSTB (2014)

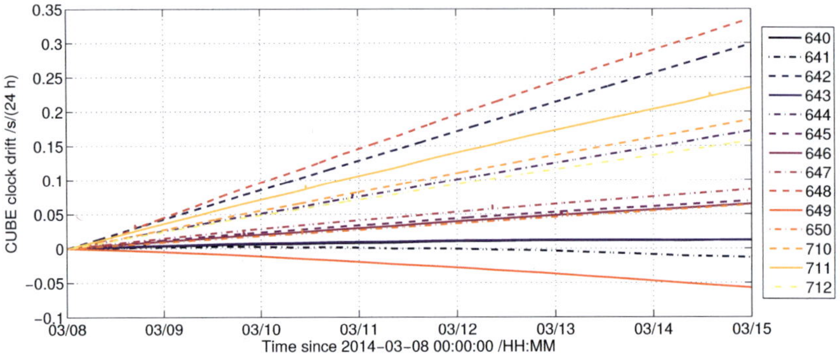

Figure 4.8: Time drift of internal clocks in CUBE data loggers during one recording week in March 2014, after the first station maintenances. All but two CUBE clocks (CUBES 641 and 649) run slower than the GPS time.

Every 30 minutes when GPS time information is available, the time vector $\boldsymbol{t}_{GPS}$ is corrected for the new GPS time information. As a result the vector is not evenly distributed any more and needs resampling to 200 Hz to distribute the new timing information consistently. However, when having unreasonable GPS time tags even resampled time vectors reveal the timing inaccuracies, as can be seen for several CUBE loggers in Figure 4.9. The clock drift vectors of CUBE 641 and 648 show positive leaps of 0.005 s forth and back between $18:00$ to $18:30$ UTC and $16:30$ to $17:00$ UTC, respectively. The built-in oscillator in the data loggers is unable to produce positive or negative leaps of this magnitude, therefore the GPS time signal must be erroneous. The GPS timing is switched on for a period of five minutes, but nevertheless the time inaccuracies are $5E+06$ times larger than the expected time transfer domain accuracy of GPS. Furthermore, the time leap exists only for one time tag, 30 minutes later, coming along with the next tag, the GPS time leaps back to the expected level. Considering that the region was affected by military activity and the experimental site was close to military stations, it is reasonable to expect that jamming and deception of GPS caused the time leaps in our GPS signals. Stud-

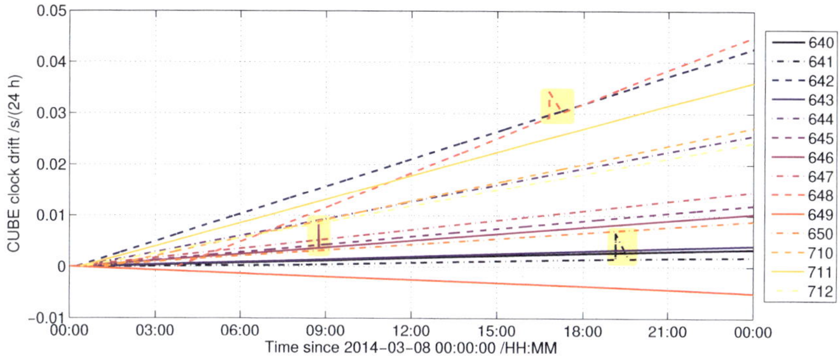

Figure 4.9: Time drift of internal clocks in CUBE data loggers during one recording day. CUBE 648 faces a maximum time drift of 0.04 s within 24 hours time delay of the internal clock compared to GPS time.

ies by Sahmoudi and Moeness (2009) and Hu and Wei (2009) confirm common GPS deception and describe impact and effect of methods of GPS deception. In the scope of this work, it is not possible to account for the erroneous GPS signals, for example by ignoring erroneous tags, since the software used to convert the data would have to be adopted for this special case by its authors. On an absolute scale the time shift of 0.005 s just equals the sampling rate of the data and therefore it still ist within the required accuracy. However, time inaccuracies cause relative time shifts at individual stations compared to the other stations which is important for array processing. Further, the resampling of the data with positive and negative time leaps, between otherwise evenly distributed data points, can cause discontinuities at the edges that will affect data perceivably in the frequency domain and cause artefacts with high amplitudes (explanations on filtering in Chapter 5.1).

4.1.4 Anthropogenic Noise

In the scope of this study, the power spectral density of ground motion velocity is one of the main subjects to the investigations (Chapters 6 and 7), in particu-

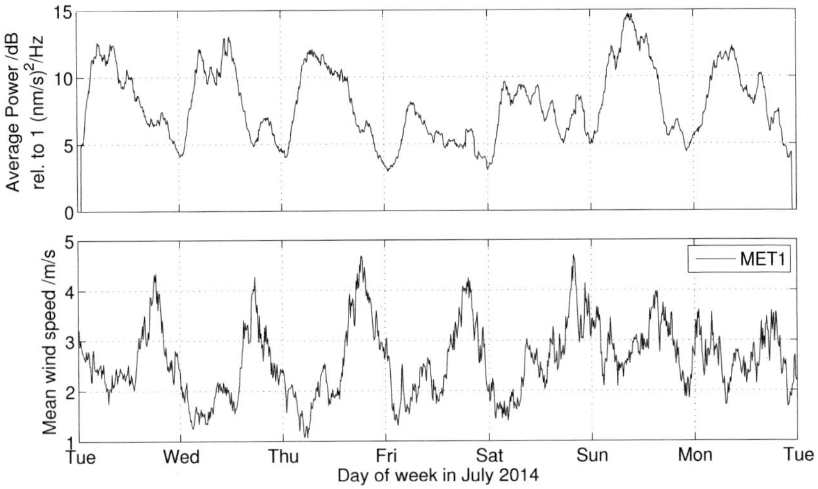

Figure 4.10: Weekly averaged power time series of DES stations in July 2014 for frequencies between 1 Hz and 100 Hz. **Top:** Power time series, averaged over all DES stations, Z-component, subsequently averaged over weekly cycle. **Bottom:** Mean wind speed at MET1 with respect to the same time axis, likewise averaged over weekly cycle.

lar the comparison of ground motion velocity during specific quiet periods with periods perturbed by the presence of wind. During calm wind the power spectral density of the ground motion velocity at the DES stations is well between the NLNM and the new high noise model (NHNM) (Figure 7.17). The overall quality of the data from the DES array, assessed by calculating PSD spectra as an indicator for the noise level, is discussed in the scope of the results, for example in Figures 6.21 and 7.2 and therefore, a detailed discussion is omitted in this section.
However, it is important to consider anthropogenic noise as factor that should not be neglected. Despite the scarce population in the Dead Sea area, roads, such as the frequented Dead Sea highway, close to the DES array are significant sources of noise in the seismological records (Figure 4.2). Figure 4.10 illustrates the average power of the DES array in July 2014 averaged as weekly

cycle. The average power is calculated according to Chapter 5.3 for the frequency band from 1 Hz to 100 Hz, where the anthropogenic influence can be expected to be dominant (McNamara et al., 2009). The graph is subjected to pronounced diurnal variations with a total span of more than 10 dB. The high peaks during day time from about 02 UTC to 14 UTC can be assigned to anthropogenic noise, considered in Chapters 7.1.2 and 7.3 as interfering factor. The lower peaks on Fridays and Saturdays, holidays in Jordan, confirm the assignment of anthropogenic factors as source for the elevated power. Comparing the power of the ground motion velocity to the mean horizontal wind speed for the same period, obtained at meteorological station MET1 (Section 4.2), excludes the wind as source for the high peaks in the power time series. Instead, smaller peaks in the power of the ground motion velocity in the evenings of each day temporally coincide with the diurnal wind speed maxima.
Using the course of the power time series of the ground motion velocity, the temporal occurrence of anthropogenic activity can be determined to reduce the seismological dataset to periods with decreased noise levels during night times (Chapter 7.1.2). The diurnal character of the course of anthropogenic noise is further used in Chapter 7.3 to reduce the anthropogenic influence on the data.

4.1.5 Array Performance

The seismological stations of the DES array were deployed following a specifically planned array design (Figure 4.2). The stations are arranged around one central station (DES01) on three different levels of distance to the centre:

1) $\approx$ 90 m: DES02, DES03, DES04, DES05, DES06

2) $\approx$ 300 m: DES07, DES08, DES09, DES10, DES11

3) $>$ 700 m: DES12, DES13, DES14, DES15

On the specific levels, the locations are selected to be evenly distributed. In radial direction from the centre, they are chosen as to avoid linear alignment of the stations. The geometry of evenly distributed, but decreasing station distances towards the centre of the array, facilitates to investigate signals in presumably broad frequency bands, independent of the propagation direction. To demonstrate the efficiency and the frequency domain in which the factual DES array is sensitive, the ATF of the station locations is calculated (cf. Chapter 2.3). In addition, known tectonic events provide the reference to conduct fk-analysis and evaluate the array performance.

The DES array was originally intended to consist of all seismological stations DES01 to DES15. Due to the low signal-to-noise ratio at DES14 (Section 4.1.4), this station is omitted in fk-analysis. Figure 4.11 illustrates the ATF of the DES array, calculated according to Equation 2.20. The fk-diagram shows the distribution of the power spectral density of the signal with respect to the wave number k and the backazimuth ϕ. The unperturbed ATF in Figure 4.11 (left) has its power maximum in the centre at $k = 0$. The width of the maximum slightly varies in different azimuthal directions. However, the performance of the array is robust for waves from different azimuthal directions.

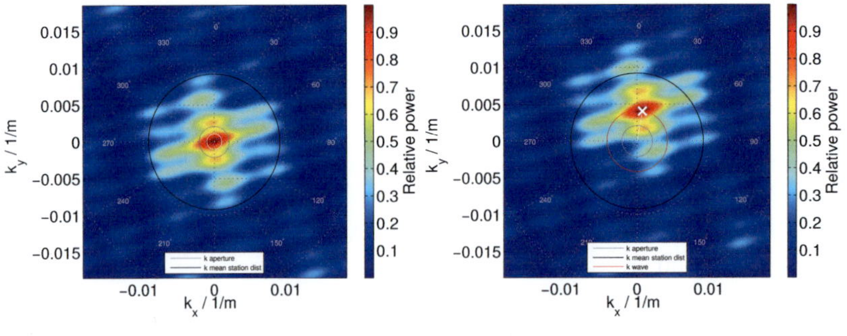

Figure 4.11: fk-diagram of array response of the DES array, without station DES14. Power is normalized with respect to the maximum of the absolute power. **Left:** Array response without incoming wave. The power maximum is in the centre. **Right:** fk-diagram with a simulated incoming wave front from 10° azimuth, a frequency of 2 Hz, and apparent velocity of 3000 m s^{-1}. The white X marks the power maximum.

The k-space (range of the wave number k) and the frequency domain are restricted by the station spacings of the array. Thereby, the minimum, mean, and maximum inter station distances dictate the k-space in which fk-analysis can be conducted. The maximum station spacing d_{max}, respectively the aperture of the array, determines the resolution defining the minimum frequency of a wave the array is able to detect (a). The minimum station spacing d_{min} determines the upper frequency limit to avoid aliasing effects when stacking the signal recordings (b):

a) Lower frequency limit: $\lambda \leq d_{min}$

b) Upper frequency limit (Nyqist frequency): $\frac{\lambda}{2} \geq d_{max}$

The restrictions of the wave length λ lead to the frequency limits for waves with an estimated velocity of $v_{app} = 3000\,\mathrm{m\,s^{-1}}$ given in Table 4.5. Illustrations of the station spacing, the station locations, apparent velocity, and frequency for which the DES array is suitable, are documented in Figure A.1, appendix.
As domain with the best response to incoming waves the limitations can be selected according to the mean station distance instead of the minimum station distance. The frequency band with the best array response thus exists between:

$$f_{min} = 1.1\text{ Hz} \; < \; f \; < \; f_{max} = 2.2\text{ Hz}.$$

The corresponding area in the fk-space is marked in Figure 4.11 between black and grey circles around the centre. Assuming an incoming mono frequent wave of 2 Hz and $3000\,\mathrm{m\,s^{-1}}$ from 10° azimuthal direction (Figure 4.11, right), the power maximum from the ATF shifts away from the centre towards higher wave numbers and closer to the limits of the mean station distance. Waves of higher frequencies would shift the maximum further towards higher wave numbers closer to the frequency limit dictated by the Nyqist frequency.
The DES array appears to be suitable for signals in frequencies between $f_{min} =$ 1.1 Hz and $f_{max} = 19$ Hz, whereas the best performance can be expected below

Table 4.5: Array specifications. d_{min}, d_{max}, d_{mean} depict the minimum, maximum and mean inter station distances, respectively. f_{min} and f_{max} represent the minimum and maximum wave frequencies suitable for fk-analysis of the DES array, assuming an apparent wave velocity of $v_{app} = 3000\,\mathrm{m\,s^{-1}}$.

Stations	d_{min} /m	d_{max} /m	d_{mean} /m	f_{min} /Hz	f_{max} /Hz
All DES stations except DES14	80	2700	700	1.1	19

$f_{max} = 2.2\,\mathrm{Hz}$. The latter defines a narrow frequency band leaving almost no variability. The broader band below $f_{max} = 19\,\mathrm{Hz}$ however, provides a more profound data base. According to the ATF the DES array is suitable to detect seismic signal independent from its azimuthal direction within the given frequency limits.

fk-Analysis of Reference Events

In November, 2014 the Comprehensive Nuclear-Test-Ban Treaty Organization (CTBTO) conducted seismic measurements between Madaba and the Dead Sea in Jordan (Gesterman et al., 2015). In the scope of the measurements, tectonic ruptures and specially induced detonations were detected and located close to the DES array. Figure 4.12 shows the locations of the events facing the location of the DES array.

To substantiate the performance of the DES array, the two smallest events, a tectonic rupture of magnitude M≈ -0.1 (event 3) and a detonation by 3 kg explosive substance (explosion 3) are subjected to fk-analysis. The information on epicentral distances, the source times, and the forces of the events, determined by Gesterman et al. (2015), can be used as reference for own calculations. Table 4.6 lists the specifications of the processed events and the results from the fk-analysis. In the following, results from the tectonic event are discussed exemplary. Corresponding results for sliding time window fk-analysis are given in Figure A.2, appendix.

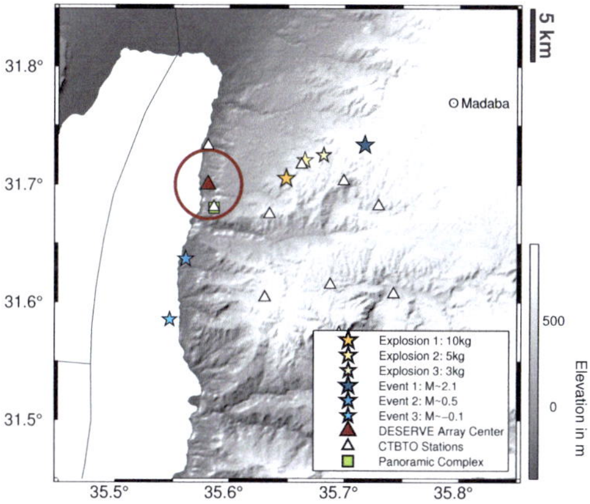

Figure 4.12: Locations of local tectonic ruptures and detonations in the scope of the CTBTO field campaign close to the DES array. Locations and magnitudes are given as presented by (Gesterman et al., 2015). The size of the detonations is given by the amount of explosive substance in kg.

Table 4.6: Specifications of two reference events and results of fk-analysis. Information on source time, epicentre, epicentral distance, backazimuth (BAZ_{ctbto}) from CTBTO test measurements (Gesterman et al., 2015). The backazimuth BAZ_{fk} and slowness Slown. is derived by fk-analysis of the DES array. Inspected events are one tectonic rupture, and one detonation in the scope of the CTBTO test measurements.

Event	**Source time**	**Epicentre**	**Epicentral distance /km**	$\mathbf{BAZ}_{ctbto}$ **/°**	$\mathbf{BAZ}_{fk}$ **/°**	**Slown. /s km^{-1}**
Tectonic event	27.11.2014,	31.5353E	13.1	195	230	0.22
M≈ −0.1	23:52:45 UTC	35.5470N				
Explosion 3, 3 kg	17.11.2014,	31.6750E	9.9	74	80	0.27
CTBTO	12:10:00 UTC	35.6814N				

The small tectonic event in roughly 13 km distance to the centre of the DES array (station DES15) can be identified in the separate records of ground motion velocity at the DES array on 27.11.2014, 23:52:50 UTC (Figure 4.13). Due to the close vicinity of the earthquake it is clearly visible in the recordings despite the small magnitude. Still, the considerable signal amplitude confirms the quality of the recordings of the DES stations.

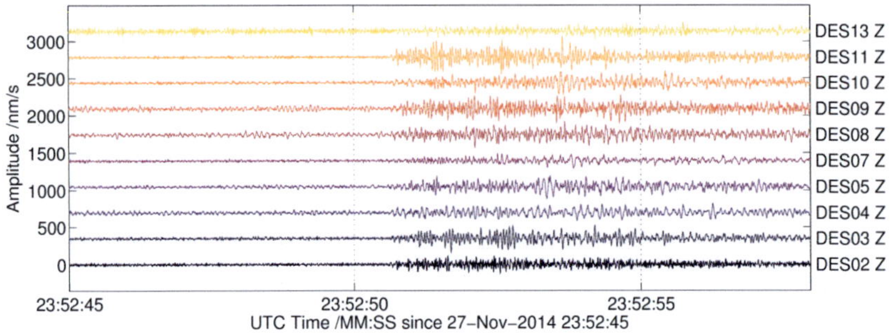

Figure 4.13: ground motion velocity at the DES array during the presence of a M=-0.1 earthquake. Applied filter: HP with $f_c = 1$ Hz.

To identify the backazimuth of the earthquake, fk-analysis is conducted for a series of consecutive, sliding time windows producing time series of power, backazimuth and slowness varying with the time series of the array beam. To distinguish different signal phases, such as p- and s-wave released by the earthquake, time windows of adequate lengths have to be chosen. In the case of a local earthquake and an array of an aperture like the DES array, a probably good time window closely embracing a single phase would be about 0.5 s. However, for frequencies down to 2 Hz the data base in one window for fk-analysis would be too sparse, in this example: only one data point. To obtain stable results, therefore longer time windows of 2 s are chosen, accepting the superposition of phases within one time window. The distinction of p- and s-phase is not possible without impairment.

Figure 4.14 shows the results of the fk-analysis for sliding time windows for the tectonic event. Note, that the increase of the power and the timeseries of slowness and backazimuth (Figure 4.14a-c) advance the arrival of the signal in the array beam (Figure 4.14d) by 2 s due to the time window length of 2 s. The absolute power of the recordings of the DES array rises from 15 dB to nearly 30 dB at its maximum as the signal of the earthquake passes (marked by grey background colour). Simultaneously, backazimuth and slowness take constant values contrary to the scattered distribution before and after the arrival of the signal. The backazimuth of about $\phi = 230°$ localizes the earthquake southwest of the array centre. The slowness of about $0.2\,\mathrm{s\,km^{-1}}$ corresponds to slowness values expected for body waves at the Dead Sea Transform fault (Ryberg et al., 2007).

The findings from the fk-analysis verify the performance of the DES array. The slowness values are in the expected range, and the backazimuth of $\phi = 230°$ is in the same dimension as the backazimuth of $\phi = 195°$ determined by Gesterman et al. (2015). The deviation of the backazimuths of 35° may be ascribed not only to location inaccuracies of the DES array, but also on potential inaccuracies of location methods based on the CTBTO array conducted by Gesterman et al. (2015). A more firm reference than the location of the earthquake is the location of the detonation in Table 4.6. As it is an induced detonation, the actual hypocentre is known. Sliding time window fk-analysis of the DES array on the explosion results in a backazimuth of $\phi = 80°$ (Figure A.2, appendix). The deviation beteen the factual backazimuth of $\phi = 74°$ and the calculated backazimuth reduces to 6°.

The results indicate the quality of the DES array. Despite restricted frequency bands and the rugged terrain, the DES array is able to assign backazimuth and slowness parameters to phases of incoming waves related to seismic events. The 6° inaccuracy of the determination of the backazimuth may be related to the strong topography. The different elevations of the station sites modify travel times of signals between the sites, requiring elevation corrections of the calculations to improve the performance (Edwards and Green, 2012).

In summary, the array performs predominantly satisfying. The ATF of the DES array shows a symmetric main maximum in the centre with only weak adjacent minor maxima. The results of fk-analysis lie within the expectancy range and the single recording stations provide records of ground motion velocity with good signal-to-noise ratios, capable to record smallest tectonic ruptures.

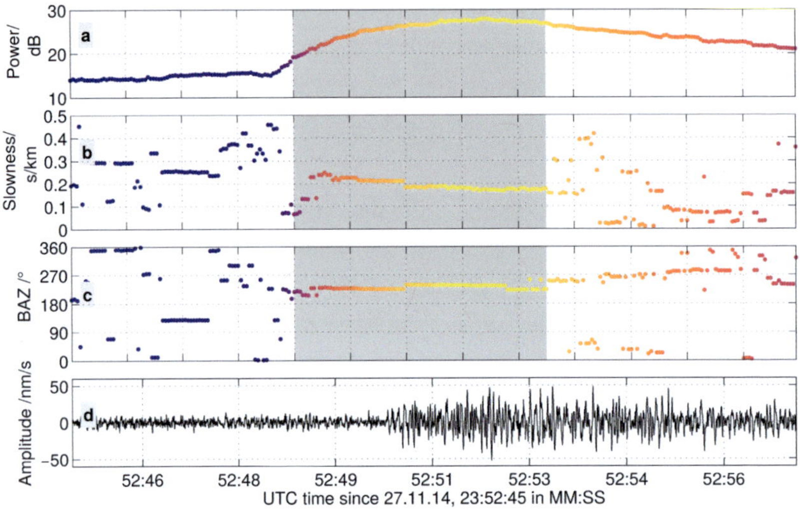

Figure 4.14: Sliding time window fk-analysis of all DES stations except DES14. Length of sliding time window: 2 s, Time steps: 0.05 s, frequency domain: 2 Hz - 18 Hz. All panels are displayed with respect to the same time axis. Colours of the dots scale with the power in Panel a. **a:** Absolute power. **b:** Slowness. **c:** Backazimuth **d:** Ground motion velocity, stacked and time-shifted according to Equations 2.13 to 2.14 (delay and sum).

4.2 Meteorological Measurements

The goal of this study is to identify meteorological information in seismological datasets and to investigate the sensitivity of seismological data to the impact of wind. Therefore, meteorological measurements are, besides the seismological measurements, the second foundation of this study.
Most important are information on wind directly at the site of the seismological measurements, provided by the meteorological Station MET1. To the west of MET1 another meteorological station, EBS3, is installed at similar latitude but at the Israel Dead Sea coast (Figure 4.15). Of all meteorological stations (Figure 3.3 in Chapter 3.2) MET1 and EBS3 are the stations whose data is

used in this study due to the vicinity of the seismological measurements. Basic parameters of the stations are listed in Table 4.7.

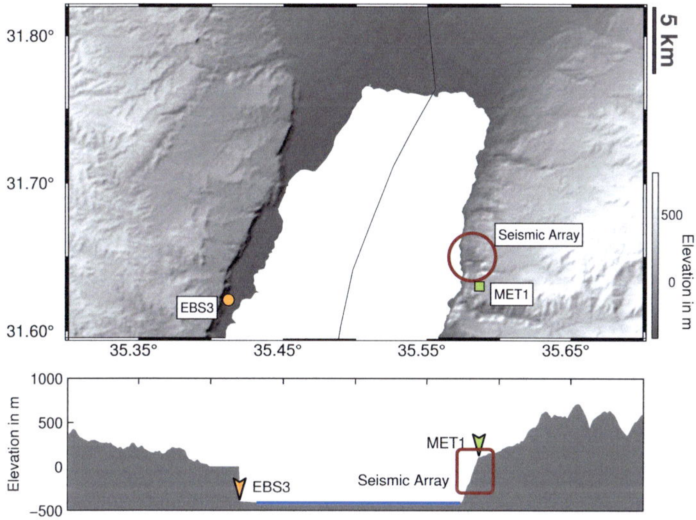

Figure 4.15: Locations of meteorological Stations MET1 and EBS3, and the area of seismic array (red). Top panel: map of the northern Dead Sea. Bottom panel: cross section of topography from east to west at latitude of MET1 (31.630 °N).

Table 4.7: Parameters of wind measurements at meteorological stations MET1 and EBS3

Identifier	Location	Coordinates	Start date	Wind measurements
MET1	Panoramic Complex, Jordan	31.63028°N 35.58616°E Alt. 128 m	24.02.2014	3-components, 10 m above ground 20 Hz sampling rate, Gill WindMaster Ultrasonic anemometer
EBS3	Kane, Israel	31.621035°N 35.412055°E Alt. −380 m	21.02.2014	3-components, 6 m above ground 20 Hz sampling rate, IRGASON sonic anemometer-thermometer

4.2.1 MET1

Station MET1 is deployed directly within the guarded area of the Panoramic Complex, about 150 m away from the buildings. It is installed on a plain area without near-by obstacles. The station includes instruments that measure multiple meteorological parameters, such as wind speed at 10 m above ground, temperature and humidity at 2 m height, air pressure, precipitation, and solar radiation. A complete list of parameters is listed in Scripture A.28 in the appendix. Most important for this work is the wind speed, therefore MET1 is equipped with an ultrasonic wind anemometer that records 3 components of wind speed (u, v, and w direction) 10 m above ground. To account also for turbulences MET1 records data with a 20 Hz sampling rate. MET1 is located at the edge of the Dead Sea valley in less than 1 km linear distance to the Dead Sea and roughly 550 m above the level of its surface. The area where the site is located forms, compared to the average slope of the valley in this region, a plain plateau rested below higher mountains in the east. Hence, MET1 is located well within the valley, but may be separated from the deeper part of the valley, where station EBS3 is located. In case of a well mixed PBL it can be expected that due to the orographic conditions MET1 experiences the same atmospheric conditions, the wind field included, as seismological Stations DES12 and DES15 located at similar elevations. However, in case of a stratified PBL DES stations in different heights may be influenced by different meteorological conditions. Observations based on data from MET1 alone might be insufficient then.

4.2.2 EBS3

Meteorological station EBS3 is located in Kane on the Israel coast of the Dead Sea. It is located very close to the Dead Sea and nearly at the elevation of the water level at −380 m. Other than MET1 station EBS3 is surrounded by high vegetation as potential obstacle. The vertical vicinity of obstacles lifts the ground surface and lowers the measurements within the logarithmic wind profile of the Prandtl Layer (Tennekes, 1973). Therefore, recordings at EBS3 might be affected by a weakened and distorted wind field.

Like Station MET1, EBS3 has multiple meteorological instruments measuring similar parameters as MET1, a detailed listing of instruments and parameters can be found in Scripture A.29 in the appendix. The wind speed is measured with 3 components obtained by an IRGASON anemometer-thermometer and recorded with a 20 Hz sampling rate 6 m above ground. Other than MET1, station EBS3 is located directly on the ground of the Dead Sea valley close to the water. Therefore, it captures winds that can be controlled by the valley's channelling effect and might be decoupled from the winds above. Another aspect to consider is that EBS3 is located on the opposite side of the valley where slope winds develop on different times of day and in opposite directions (Figure 3.2). However, EBS3 can provide information about the local situation within the valley MET1 cannot obtain. Despite the larger distance of 15 km between EBS3 and the seismological stations and its setting on the opposite side of the valley, local conditions at the lower seismological stations might be more similar to conditions at EBS3 than to MET1. In any case, data from both stations has to be taken into account to achieve a more complete image.

In this study wind data from MET1 and EBS3 is used with the original sampling rate of 20 Hz except when otherwise mentioned. In contrast, the terms mean wind speed or mean wind direction imply the 10 minute averages of the originally sampled wind data. When referring to wind speed or mean wind speed, the horizontal projection is described unless otherwise mentioned.

4.3 GEOFON Station GHAJ in Ghour Hadithe

The closest permanent seismological station operating during our campaign and providing public data is GEOFON[4] station GHAJ in Ghour Hadithe close to Al Mazraa. It is located about 40 km south of the Panoramic Complex near

[4] GEOFOrschungsNetz is a global seismological broad-band network operated by GFZ. Further information can be obtained in GEOFON Data Centre (1993)

the southernmost end of the Dead Sea in Jordan (Figure 4.1). Other than the temporary seismological stations near the Panoramic Complex GHAJ is a permanent station and therefore its installation and setup are more elaborate providing better shielding and a higher signal-to-noise ratio. Due to the station's vicinity to our campaign, similar meteorological and seismological conditions, and the higher quality, data from GHAJ is used in addition and as a reference in this work. Table 4.8 lists basic information on station GHAJ and on data streams used here.

Station GHAJ is located about 4 km apart from the Dead Sea shore but already 360 m above its water level. The area is steep, eroded, and without noteworthy vegetation, similar to the area of the lower temporary seismological stations of the array, but without sealed roads in the vicinity.

In this work data streams HHZ, HHN and HHE are used, like data from our temporary stations covering three spatial directions vertical, north and east. Data is provided with 100 Hz sampling rate, measured by a Trillium-240 broad band seismometer. Like data from our own network, raw data from GHAJ at first has to be converted from counts into physical units (Chapter 5.1). Required parameters such as the instrument gain are listed in Table A.3 in the appendix. Merely the village of Ghour Hadithe is close by and few surrounding Bedouin cultivating farmland. The seismometer at GHAJ is installed in an isolated chamber 3 m below the earth's surface and therefore also shielded against atmospheric influence, such as direct wind making GHAJ a suitable reference for results obtained from the temporary stations.

Table 4.8: Characteristics of permanent seismological station GHAJ operated by GFZ and data streams used in this work.
Source: http://geofon.gfz–potsdam.de/waveform/archive/station.php

Identifier	Location	Coordinates	Start date	Streams	Local depth	Sensor
GHAJ	Ghour Hadithe, Jordan	31.3034°N, 35.5665°E, Alt. −58 m	01.06.2011	HHZ, HHN, HHE	3 m	Trillium-240

5 Seismic Data Processing and Routines

Seismological records of ground motion velocity used in this thesis are subjected to many processing steps before entering a state where they can be used for analysis (Chapters 6 and 7). The basic processing includes several inevitable steps before applying methods for the analysis of data. The digital raw data has to be converted into a processable format, it has to be resampled, corrected for data gaps, and converted into factual physical units to enable interpretation. This chapter describes required routines for basic data processing, mathematical and physical background, and methods to process and examine data for analysis, as a base for the evaluation in subsequent chapters.
Most of the conversions and calculations, as well as all the graphic presentation of results are conducted using MATLAB. Open source programme codes from external sources for MATLAB and other algorithmic languages are indicated in the text where the method is described and listed in a nutshell in Tables A.1 and A.2 in the appendix. Important to mention for this work is a library of codes, the KaSP toolbox[1] for MATLAB. It is a collection of functions for seismological data processing, created and developed at the Geophysical Institute at KIT. All processing steps except of the conversion of seismological raw data from CUBE into MSEED format is conducted using MATLAB (Section 5.1.1). Except of some functions referenced in text and appendix, the calculations are conducted with programmes and functions from the author of this thesis. An important and specific approach are methods and calculations developed solely for this study, to collocate seismological and meteorological records.

[1] The Karlsruhe Seismology Processing Toolbox for MATLAB (KaSP) (C) 2009-2012 by the KaSP-Team. Geophysical Institute, Karlsruhe Institute of Technology

The following sections describe methods applied to data roughly in chronological order. Section 5.1 covers the general seismological data conversion and processing that is applied to all seismological records upfront conducting further calculations. Subsequent Section 5.2 describes common calculations in the frequency domain, such as the estimation of the power spectral density (PSD), needed for analysis in Chapters 6 and 7. The last Section 5.3 explains methods, developed in this study, to quantify correlations between meteorological and seismological datasets. By intention, figures in this chapter, schematically illustrating selected routines, are labelled with physical variables. But corresponding units are omitted in order to pay attention to the methods instead of potential interpretations.

A: Basic processing → **B: Calculation of Spectrograms** → **C: Correlation of data sets**

5.1 General Seismological Data Processing

A: Basic processing

All seismological data

i Conversion of raw data into MSEED and resampling
ii Removal of mean and linear trend
iii Application of taper and high-pass filtering
iv Conversion of data units into physical units

Data from DES Stations

v Removal of instrument response
vi Rotation of horizontal components

The following section serves to document fundamental processing routines as to make the results reproducible and to estimate the quality of results. All seis-

mological records used in this thesis are subjected to the routines described here before any further usage.
The routines described here, belong to seismological standard processing for untreated raw data. Anyhow, they are briefly described in this section because there exist no universal parametrizations for seismological raw data, in fact, the processing has to be adjusted to meet unique characteristics of the records and to fulfil specific research purposes. All calculations from in **ii** to **vi** are conducted using MATLAB. The required filters and functions, such as the FFT, exist for MATLAB and the KaSP toolbox. However, the full range of conversions are calculated without ready-made tools converting raw data into processable data, but manually as described below. Seismological records from DES stations are subject to all processing steps **i** to **vi**. Ground motion velocity records from the GEOFON station is publicly available in MSEED format and is subjected to processing steps **ii** to **iv**.

5.1.1 Data Conversion and Filtering

i Conversion of Raw CUBE Data into MSEED

Seismological raw data from the DES stations, available in CUBE format is converted into MSEED format using `cube_mseed` (further information in Table A.2 in the appendix). Conversion parameters are chosen as to incorporate GPS time information, resample data to a steady sampling rate, and cut files into sequential files of one hour duration each. Due to problems with the GPS reception (cf. Chapter 4.1.3) the assessment of the timing information is problematic and proven successful so far only with release 2015,07,24. Open functions for MATLAB interpret the binary format of MSEED.

Seismological records preferably exists as continuous time series. In this case due to station maintenance and GPS problems there are gaps in the time series making it impossible to transform the data into the frequency domain. When importing data into MATLAB it is therefore essential to merge sequential time segments and pad gaps in data with zeros that have to be ignored when operations in the frequency domain are conducted.

ii Removal of Mean and Linear Trend

As part of the basic processing routine data is subjected to several consecutive procedures: removal of offset and long period trend, and filtering of the data. Before filtering, the beginning and the end of the data segment are treated with a statistical Tukey window taper to flatten the amplitude of the signal towards the ends of the segment preventing discontinuities and consequential artefacts when applying filters.

iii Tapering and Application of High-Pass Filter

It is essential to apply at least minimal high-pass filtering (HPF) to the data in order to be able to remove the instrument response without artefacts (cf. Section 5.1.2). The time series are filtered using a Butterworth zero-phase of 2nd order in the time domain. The selected corner frequencies f_c of the filters are chosen depending on the characteristics of the sensor, respectively on the shape of the instrument response as illustrated in Figures 4.6 and 4.7. The corner frequencies are selected slightly below the lower declinations of the responses. For reasons of consistence, data from broad band GEOFON station GHAJ is filtered with the same corner frequency as data from our Trillium Compact sensors:

- Mark Products sensor: High-pass filter with $f_c = 0.2$ Hz
- Trillium Compact sensor: High-pass filter with $f_c = 0.005$ Hz
- Trillium-240 sensor: High-pass filter with $f_c = 0.005$ Hz

iv Conversion of Data into Physical Units

Seismometers do not quantify obtained records into physical units but in electronic pulses transmitted dimensionless in numbers of counts to the data logger. The signal that is yet processed as described in **i** to **iii** still exists with respect to electronic units of counts. Therefore, it has to be converted to physical units before interpretation, in this case to ground motion velocity in $[v]$=nm s^{-1}. The electronic signal $g(t)$ has to be multiplied by a conversion factor m_c depending on individual properties and settings of seismometer and data logger. It therefore differs for records from stations with different seismometers and loggers. In general, the conversion factor m_c can be described as in Equation 5.1:

$$m_c = \frac{AD}{\gamma \cdot s}. \tag{5.1}$$

m_c is determined by the analogue-to-digital conversion factor $[AD]$=μ V. AD is a constant that depends on the data logger, the gain γ, a specific multiplier that can be set for the data logger, and the sensitivity $[s]$=V s m^{-1} which depends on the recording component of the seismometer. Average values for AD, γ, and s are listed in Table 4.4 for the DES stations, exact values for each station can be found in the appendix in Scriptures A.17 to A.27. Table A.3 in the appendix lists the essential parameters for station GHAJ. The records accordingly can be transformed into ground motion velocity $v(t)$:

$$v(t) = y(t) \cdot m_c. \tag{5.2}$$

v Removal of Instrument Response

The resulting ground motion velocity is, however, not yet as close to the original input signal $f(t)$, as possible. It still rather corresponds to the signal output $g(t)$ of the seismometer (cf. Chapter 2.1.3), filtered and converted into physical units, however the influence of the instrument remaining. To approximate the input signal $f(t)$, subsequently the instrument response has to be removed as described in 5.1.2.

vi Rotation of Horizontal Components

As a last part of the general data processing, after having removed the instrument response, the rectangularly oriented horizontal recording components are rotated towards the geographic north to compensate the declination of the geomagnetic geomagnetic north (cf. Chapter 4.1.3). The horizontal components 1 and 2 turn to N and E. The vertical component 0 does not need to be changed, it already corresponds to Z.
Data from station GHAJ is already available in MSEED format and can be directly included in the processing routines without having to resample or correct the data for clock drifts or rotation of the components. In other respects the data is subjected to the same processing routines as data from DES stations.

5.1.2 Removal of Instrument Responses

The instrument response describes the behaviour of the sensitivity of a seismometer. When recording ground motion velocity, the seismometer behaves similar to an analogue filter to the signal. Accordingly, it affects the record and thus the output record differs from the input record. To minimize the difference between original ground motion velocity and output record from the seismometer, the filter, respectively the influence of the instrument, has to be removed from the recorded signal. Section A.1 describes the mode of operation of seismometers and the consideration of the instrument response mathematically and in more detail. Relations and equations described here, can be retrieved from Scherbaum and Schmidtke (2007) and Wielandt (2012).
As indicated by the instrument responses in Figures 4.6 and 4.7 in Chapter 4 the response of the instrument highly depends on the frequency of the recording. The physical limitations of the instrument lead to a behaviour that can be described by the instrument response.

The instrument responses of all seismometers deployed for this thesis is given as poles and zeros. The Mark Products seismometers have unique instrument responses for each of the three recording components Z, N and E (see Appendix A.17 to A.27), thus, for all components the instrument response has to be calculated and removed separately. By contrast, the calculations of the instrument responses of the Trillium Compact instruments is simpler, since there is only one instrument response determining all channels of all instruments of that type.

The instrument response exists in notification of poles and zeros and a normalization factor. To work with it, the response has to be transformed into a processable format. From the z-domain it is converted into a function of frequency $H(\omega)$, as presented in Figures 4.6 and 4.7.

In the time domain the output signal $g(t)$ corresponds to the convolution of response and input signal $f(t)$. In the frequency domain, the relation is expressed more fundamental as multiplication of response and input signal. Therefore, in the frequency domain the removal of the response from the given output signal is quite feasible:

$$F(\omega) = \frac{G(\omega)}{H(\omega) \cdot c}, \tag{5.3}$$

with $F(\omega)$ and $G(\omega)$ as the Fourier transforms of $f(t)$ and $g(t)$. c represents the specific normalization factor coming along with instrument response (Table 4.4).

The available output signal $g(t)$, already processed according to the descriptions in Section 5.1.1, is transformed to $G(\omega)$ using Fast Fourier transform (FFT). It is then divided by the transformed instrument response including the normalization factor. The inverse FFT (iFFT) transforms the signal $F(\omega)$ back into the time domain to receive the input signal $f(t)$, adjusted for the influence of the instrument:

$$f(t) = \text{iFFT}\left[\frac{G(\omega)}{H(\omega) \cdot c}\right]. \tag{5.4}$$

In the time domain, the record $f(t)$ is filtered again using the same high-pass filter as prior to the removal of the response. The records are then available as time series of seismic ground motion velocity and base for all further calculations.

5.2 Seismological Data in the Frequency Domain

B: Calculation of spectrograms

- Spectrograms of ground motion velocity
- Power spectra of ground motion velocity
- Power time series of ground motion velocity

Seismological records at hand, processed as described in Chapter 5.1, are time series of velocity in units of $\mathrm{nm\,s^{-1}}$. However, a major part of the analysis in this study is conducted in the frequency domain facilitating to interpret data in terms of power spectral density instead. When regarding time series, one can read the absolute amplitudes, polarization, and exact time information, within the timing or amplitude accuracy the filters applied to the data allow for. However, when comparing it to meteorological data, time information in high-resolution, such as given for the seismological records with a sampling rate of 200 Hz, is not necessary, since an average wind speed with a sampling rate of 10 min is common to describe changes in the local wind field. The following sections describe the basic methods used in this study to process, combine, and compare ground motion velocity and wind speed records.

5.2.1 Power Spectral Density of Ground Motion Velocity

The PSD is a useful tool to estimate the impact of signals in seismological records with respect to the frequency domain. In this study, the PSD of a signal y is estimated using single sided FFT:

$$P = 10 \cdot \log_{10} \frac{2}{f_s \cdot N} \cdot \mathrm{FFT}(y)^2. \tag{5.5}$$

Thereby, N is the number of data samples, f_s the sampling rate in Hz, and $\mathrm{FFT}(y)$ is the fast fourier transform of the signal y. The FFT transforms the signal into the frequency domain and such is P a function of the frequency ranging from the inverse signal length to the Nyqist frequency $f_s/2$. P is calculated as a logarithmic parameter and therefore the PSD of the ground motion velocity $[v]$=nm s^{-1} can be described in units of $[P]$=$10 \cdot \log_{10}(\mathrm{nm}^2\,\mathrm{s}^{-2}\,\mathrm{Hz}^{-1})$. In this study, the PSD is usually given with respect to the signal maximum or to $[P]$=1 $\mathrm{nm}^2\,\mathrm{s}^{-2}\,\mathrm{Hz}^{-1}$ in $[P]$=dB.
The calculation of the PSD as described here is used in this thesis to gain information on seismological records of periods up to one day. It transforms records of multiple seismological stations and components into a comparable shape to investigate specific characteristics in the frequency domain.

5.2.2 Spectrograms

Spectrograms of ground motion velocity in this thesis are used to get a detailed image of ground motion velocity for short periods such as a few hours but also for long periods such as several days or a few months. This study uses spectrograms for two purposes:

- Qualitative interpretation of the spectrograms itself for selected stations and case studies (Chapter 6)
- Essential data base for further calculations (Chapter 7)

Table 5.1: Settings of spectrogram calculations

Identifier	**Frequency band /Hz**	**Window length /s**	**Frequency resolution /Hz**	**Stations**
f1	0.05 - 0.1	1500	0.005	DES01,12-15
f2	0.1 - 1	1500	0.01	DES01-15
f3	1 - 100	360	1	DES01-15
	1 - 50	360	1	GHAJ

While spectrograms provide an estimation of the power spectral density over frequency on a time axis, the complex images make it hard to investigate records of multiple seismological stations and components in comparison. For more systematic studies, however, the spectrograms provide an extensive data base for further calculations.

Spectrograms are calculated by concatenating spectra of consecutive time windows. In this study, an adaptive multitaper method (Percival and Walden, 1993) is used[2]. The length of the consecutive time windows for the spectral calculations varies. It depends on the intended frequency band of the spectrogram and the purpose. In this study, separate spectrograms are calculated for three adjacent frequency bands $f1$, $f2$, and $f3$ between 0.05 Hz and 100 Hz, listed in Table 5.1. Depending on the chosen frequency domain, the time window length has to be adjusted to ensure that enough data points remain in each time window for the calculations. Spectrograms are calculated for all frequency domains $f1$, $f2$, and $f3$ for the broad band stations. The short period stations loose sensitivity with decreasing frequencies below 1 Hz, therefore the short period data is not processed for domain $f1$. Spectrograms for Station GHAJ have an upper frequency limit of 50 Hz due to the sampling rate of s_r=100 s.

Figure 5.1 provides a schematic example of a spectrogram as calculated in this study. It contains gridded PSD data with respect to a time and a frequency

[2] MATLAB code based on function `longtermspectrogram.m` from KaSP Toolbox; cf. Table A.1

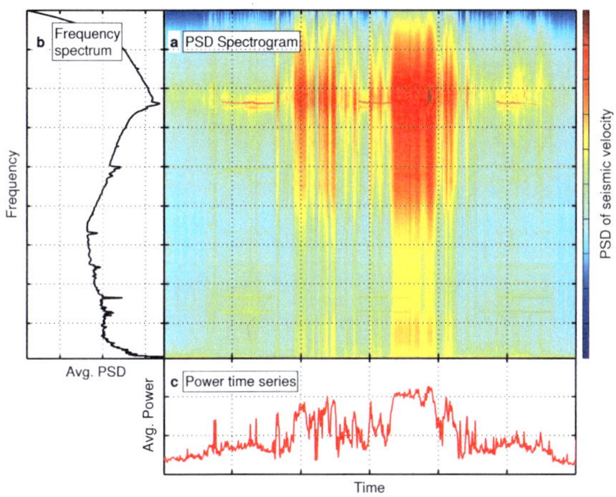

Figure 5.1: Schematic illustration of PSD spectrogram, derived frequency spectrum, and power time series. **a:** PSD spectrogram with respect to time and frequency axis. Colours indicate the PSD. **b:** Derived frequency spectrum with respect to the same frequency axis. **c:** Power time series with respect to the same time axis.

axis. Besides allowing the qualitative interpretation of the PSD with respect to frequency and temporal occurrence, the spectrogram dataset is the base for further analysis, as described in the following section. In addition to the extensive data matrices, averaged spectrogram data, taken for the frequency band (Figure 5.1b), or the time axis (Figure 5.1c), are used in this study. Averaging the data over all time windows (b) results in spectra comparable to the spectra obtained via FFT (cf. Equation 5.5). It is less accurate, but still convenient when analysing long periods of multiple days, where FFT becomes too memory-consuming. Frequency spectra of three month length with respect to the wind speed (Section 5.3.1, Chapter 7.1.1) are derived from spectrogram data.
Another important tool in this study are power time series (Figure 5.1c), likewise derived from the spectrograms by averaging over the frequency domain instead of the time. The power time series are a simplification of the spectra disregarding fluctuations in specific frequencies, anyhow they are measures for the temporal change of power within a broader frequency band at hand. In this study, the course of the power time series is compared to the course of wind speed over time by calculating cross-correlations as a measure for the correlation of the datasets (cf. Section 5.2.3, Chapters 6.3.1 and 7.3).

5.2.3 Quantification of PSD of Ground Motion Velocity and Wind Speed via Cross-Correlations

Chapters 6.3.1 and 7.3 use time series of average power of ground motion velocities in order to compare seismological records with the course of changing wind speed. Power time series, as illustrated in Figure 5.2c are calculated from the spectrogram data of all seismological stations. The courses of the power time series and the wind speed can be compared quantitatively by using normalized cross-correlation estimates (Bendat and Piersol, 2010) of the normalized traces. Resulting cross-correlation coefficients between -1 and 1 range from anti-correlation (-1) over no correlation (0) towards total correlation (1) of the datasets.
Chapters 6.3.1 and 7.3 use the cross-correlations as instrument to estimate the influence of local wind changes to the seismological ground motion velocity. To restrict the wind-induced signals in the seismological records preferably to the local wind, the power of the ground motion velocities assigned with the local wind speed is confined for the frequency domain $f3$ only.
For the case studies in Chapter 6.3.1, cross-correlations between wind speed and power time series are calculated for the whole time periods of the events displayed. For the whole study period of three months in Chapter 7.3, cross-correlations of the whole period become too insensitive due to the lengths of the signals. As reasonable time period to compare the change of wind speed to the ground motion velocity, signal lengths of one day are chosen instead, in accordance to the wind speed spectrum by Van der Hoven (1957).

5.3 Collocating Seismological Records with Wind Speed

Seismological and meteorological records usually are not intended to be compared, therefore, they don't come as a collection in one interrelated dataset to investigate. The difficulty in pooling the data is not only to bring the data to a common base, such as comparable time vectors, allowing to index and assign

C: Correlation of data sets

- Pooling of meteorological time series and ground motion velocity spectrogram data
- Conversion of sampling rates according to frequency domain
- Wind dependent spectra
- 2D Histograms of PSD of ground motion velocity
- Ground motion susceptibility
- Composition of wind speed, wind direction and spectrogram data

both datasets. The problem is also that the ground motion velocity, like the meteorological records available as time series, is not straightforward to compare quantitatively to wind speed records. In this thesis, a possibility to pool and interpret seismological and meteorological records as a common dataset, is developed. The dataset, in the following referenced to as meteorological-seismological (MS) dataset, forms the base for all essential calculations in this thesis. It comprises mean horizontal wind speed and wind direction time series on the one hand, spectrogram data of the PSD of ground motion velocities on the other hand.

To clarify the motive for using spectrogram data instead of original ground motion velocity time series, Figure 5.2 shows an example of a dataset of seismological records, including a time series of ground motion velocity (a) and corresponding data of power spectral densities (b,c), and average wind speed (d) with respect to the same time axis.
The example comprises a period of changing wind speed and shows the influence on ground motion velocity. The amplitudes of the ground motion velocity correlate with the course of wind speed change. High wind speed comes along with high amplitudes. However, the signals are not directly comparable, and from the seismic time series only no conclusions on potential frequency changes can be assumed. The spectrogram (Figure 5.2b), calculated

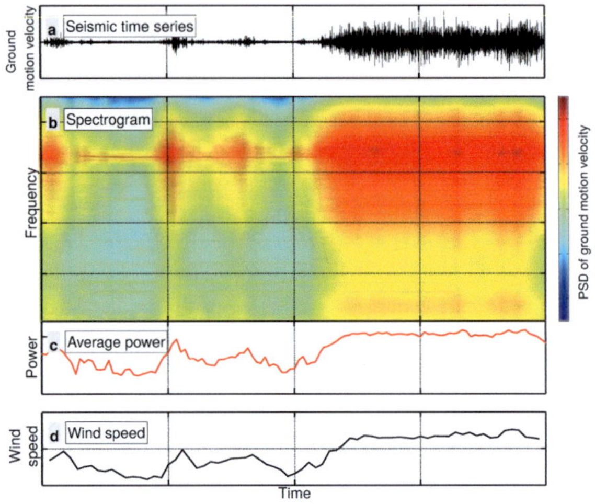

Figure 5.2: Schematic example of a meteorological-seismological dataset. Data sample is from Stations DES15, Z-component, and MET1. **a:** Time series of ground motion velocity. **b:** Spectrogram of PSD. **c:** Average power based on spectrogram data. **d:** Mean wind speed.

from the ground motion velocity time series (a), provides information on both frequency and time domain, though with reduced resolution. Averaging the PSD from spectrogram by the frequency yields a time series of average power of ground motion velocity(c). Other than the original time series of ground motion velocity the course of the average power from the spectrogram scales with the wind speed record (d) and thus, can be compared more straight-forward, for example by using cross-correlations.

The reduced temporal resolution of the spectrogram data suffices to consider non-turbulent changes in the wind field whilst providing information depending on specific frequencies. Ground motion spectrograms and records of wind speed and wind direction from MET1, merged to MS datasets, form the base dataset for this study. Therefore, spectrograms are calculated continuously for all seismological stations, including station GHAJ, all recording components, and for three frequency domains (cf. Table 5.1) separately. Meteorological time series of wind speed and wind direction are individually decimated, based on floating mean averaging, down to the temporal resolution of the spectrogram data in order to form a unity with respect to the same time axis. The resulting MS dataset can be treated as one general dataset and allows to directly associate changes of the wind at the meteorological station MET1 with PSD of ground motion velocity. The common time axis thereby is the key to the comparison of the datasets. It is possible to filter the complete set for certain wind speed levels or wind directions by associating the complied conditions to certain time samples within the study period, ascribing it to according spectrogram samples.

5.3.1 Wind Speed Dependent PSD Spectra

The data base for the calculation of spectra of ground motion velocities with respect to wind speed in Chapter 7.1.1 is taken from the MS dataset. Dividing the mean horizontal wind speed into u_h:

$$0 \leq u_h < u_{h,max}); \qquad u_h \in \mathbb{N} \tag{5.6}$$

results in an even wind speed grid with a resolution of $1\,\mathrm{m\,s^{-1}}$. By assigning the wind speed time series to it, it is possible to sustain the corresponding time stamps into each grid point as well. Going along, the data points from the whole MS dataset can be collocated to the grid as well allowing to assign the ground motion spectrogram data to specific wind speed steps u_h. Analogue to the wind speed grid, the dataset can be assigned to a grid of wind directions as well (Section 5.3.4).

For each wind speed grid a separate MS dataset then exists. Figure 5.3 illustrates the selection of data into one exemplary wind speed bin (blue frames). The arrows point towards the corresponding time information. Using the timing information, data from the spectrogram is assigned, keeping the existing frequency information of the spectrogram.

To obtain PSD spectra for specific wind speed bins, for each bin the spectrogram data (blue frames in spectrogram) is averaged over the selected time iterations. The result is each one specific PSD spectra for each wind speed bin (Figure 5.3, left). Calculating the spectra from spectrograms of all frequencies and averaging for all stations results in wind speed dependent spectra for the seismological array, characterizing frequency-specific excitation of ground motion velocity with wind speed as shown in Figure 7.2.

5.3.2 Relation between PSD of Ground Motion Velocity and Wind Speed

Chapter 7 establishes a relation describing the power spectral density of the ground motion velocity as a function of the horizontal wind speed (cf. Equation 7.1). The calculations are based on the MS dataset, initially without considering different wind directions (Chapter 7.1). However, Chapter 7.2 distinguishes records assigned to consecutive segments of wind directions.

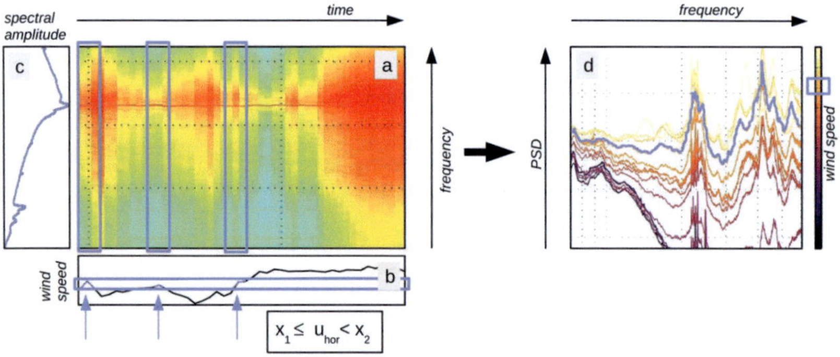

Figure 5.3: Left: Schematic example of the collocation of the MS dataset into one exemplary wind speed bin $x_1 \leq u_h < x_2$. The blue frame in **b** marks the wind speed bin illustrating the intersection with the wind speed time series. The arrows mark the time stamps of the assigned data points. Blue frames in the spectrogram (**a**) refer to corresponding data that are averaged for each frequency to obtain the spectra (**b**) for the given wind speed bin. **Right:** Resulting frequency spectra for each wind speed (**d**). The blue graph corresponds to the wind speed marked in the colour bar and stands exemplary for the selected wind speed in **b**.

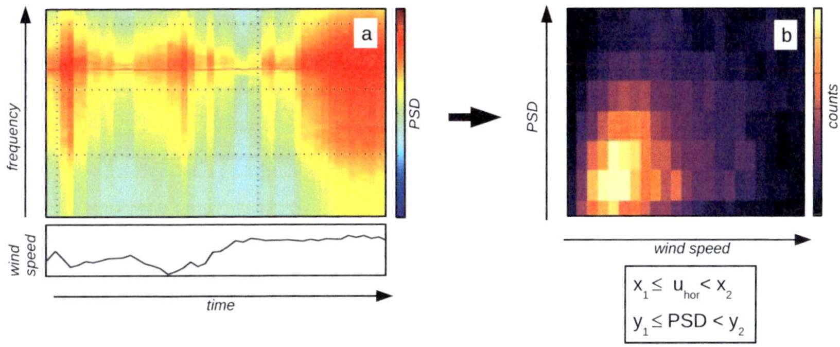

Figure 5.4: Schematic illustration of 2D histograms of PSD of ground motion velocity versus wind speed. **a:** Spectrogram of PSD and wind speed time series. **b:** Resulting 2D histogram.

Describing the calculations without respect to certain wind directions, the spectrogram and wind speed data form a two-dimensional grid of wind speed and PSD bins:

$$0 \leq u_h < u_{h,max},$$

$$\min(PSD) \leq PSD < \max(PSD),$$

with increments of $\mathrm{d}u = 0.2\,\mathrm{m\,s^{-1}}$ and $\mathrm{d}PSD = 1\,\mathrm{dB}$. Analogue to Figure 5.3 the spectrogram data points are assigned to the specified wind speed bins, and subsequently assigned to the PSD bins. Entries in each bin are counted as histogram. The resulting 2D histogram then exhibits a dependency of PSD on wind speed on a grid. Thereby, colours indicate the number of data points included (Figure 5.4).

The type of diagram is used to display the spectrogram data with respect to wind speed without needing to show all data points in one graph that would strongly overlap for long periods such as three months. The data as displayed here, however, is not subject to further calculations in the present, averaged shape. For calculations of the ground motion susceptibility, for example, data is used as described in Section 5.3.3.

5.3.3 Ground Motion Susceptibility

The ground motion susceptibility S (Equation 7.1) is a model developed in this study to quantify the vulnerability of ground motion velocity to the friction-induced effects of wind. The lower threshold of PSD data observed in the 2D histograms (Figures 5.4 and 7.3), scaling with the horizontal wind speed, is assessed by the 5%-percentiles of the PSD in each defined wind speed bin of $du = 0.2\,\mathrm{m\,s^{-1}}$ (cf. Section 5.3.2). Linear regressions conducted for the 5%-percentiles are congruent.

The ground motion susceptibility is calculated for all seismological stations and recording components, with the same settings. Thereby, the susceptibility is calculated for the frequency domains $f1$ to $f3$ (Table 5.1), based on the full MS dataset and on a selection of multiple segments of wind direction (Chapter 7.2).

5.3.4 Dependency of Seismic Records on Wind Direction

The 2D histograms illustrate the dependency of the PSD of ground motion velocity on the mean horizontal wind speed. However, it is not possible to include information on the wind direction due to the averaging of the data into the 2D grid of PSD and wind speed. Instead, each point of the MS dataset needs to be considered without a grid.

Assigning each data point from the spectrogram data (Figure 5.5a) to the corresponding exact wind speed and wind direction values and plotting each data triplet of PSD, wind speed and wind direction, results in a graph as illustrated in Figure 5.5b. The PSD of the ground motion velocity is displayed over the wind speed, similar to the 2D histograms, but each original data point is included in the illustration. The information on wind direction remains and can be displayed according to a colour scale. The main disadvantage of this kind of illustration is that long periods of several weeks provide large amounts of data that can not be displayed without ambiguity. The data points overlap concealing significant amounts of information especially in dense sectors.

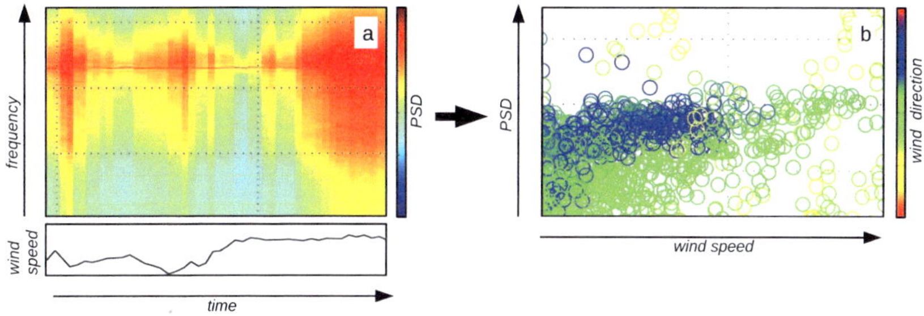

Figure 5.5: Schematic illustration PSD of ground motion velocity versus wind speed with colour indicating the wind direction. **a:** Spectrogram of PSD, wind speed and wind direction time series. **b:** Reorganized PSD data depending on wind speed and wind direction (colour).

To avoid the issue of overlapping data points and to lay focus on the distribution of PSD data with respect to the combination of wind speed and wind direction (Chapter 7.2.2, i.e. Figure 7.11), a different arrangement of illustration is appropriate. The PSD is illustrated on a colour scale with respect to wind speed and wind directions. The arrangement allows interpolating the PSD data points on a fine meshed grid of wind speed and wind direction. Further, rotating the data into an azimuth coordinate system, produces a geometrically descriptive illustration of the PSD, indicated by colours depending on the geometry of the wind field (Figure 5.6). Thereby, the azimuth defines the wind direction and the radial component defines the wind speed.

Reduction of Anthropogenic Noise

Benefiting from the array configuration of the seismological stations the power time series can be stacked and averaged without normalization. The stacking especially becomes interesting when investigating the correlation of seismological records for long periods, where the stacked power time series describe the sensitivity of the whole array to the overall local wind speed changes (Chapter 6.3.1).

However, the stacked power time series for long periods reveal a strong diurnal change of power, confirmed by Figure 4.10, that can be assigned to anthropogenic noise. The stacking of the signals reduces site-specific, signals from individual stations, by simultaneously amplifying common signals at multiple stations. Hence, not only wind-related signals are amplified, but also the anthropogenic noise. Due to the diurnal reoccurrence it is possible to calculate the anthropogenic ground noise as average diurnal cycle for the stacked data. The influence of the anthropogenic noise thus can be minimized by dividing the stacked power by the noise signal (Chapter 7.3).

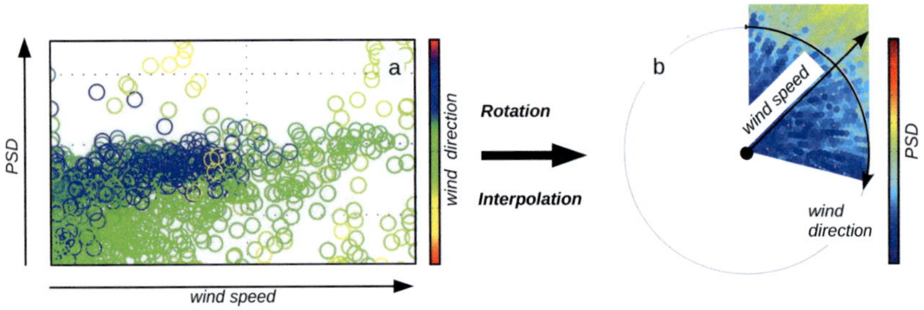

Figure 5.6: Schematic illustration of PSD of ground motion velocity with respect to wind speed and wind direction interpolated on azimuth coordinates. **a:** Ground motion PSD versus wind speed, including wind direction on coloured scale. **b:** Data triplets reorganized as wind speed versus wind direction, transformed into azimuth coordinates. PSD is indicated by colours.

6 Impact of Severe Wind Events on Seismological Records

It is well investigated that during severe wind the noise level of seismological records can be increased (Mucciarelli et al., 2005). However, the amount of studies on the research topic of wind influencing seismological recordings is sparse and the magnitude and the kind of impact may vary individually due to site effects and the type and the seismic insulation of the seismological stations. To assess first insight into how wind affects the seismological measurements conducted in this study, it is likely to start investigating events with presumably high wind speed.

High local wind speed can be caused by multiple reasons ranging from local scale, such as slope winds, to synoptic scale, such as storms (Holub et al., 2009). Ground motion is not only influenced by local changes in the atmosphere but also by more severe impacts in long distances. Storms with different properties, such as strength, distance to the pressure minimum, duration or genesis influence seismological records in various ways. There are several questions that can be addressed by investigating selected events with the highest wind speed within the measurement period. Likely, the first matter is to find out if the data reveal any correspondence between wind speed and signals in seismological records at all. If there is an influence of wind speed on seismological records: what does it look like and which frequencies in the seismological records are most sensitive to wind? Is it possible to link between the origin of the meteorological event and the pattern in the seismological records?

By reducing the whole dataset of 11 months to a limited set of case studies it is possible to look at the time series in more detail and to carve out differences

in the signals recorded at the seismological stations. Comparing the seismological records at different stations can indicate site effects and reveal various quality levels of the stations.
When examining single meteorological events it is also possible to investigate type and genesis of the event in detail. Therefore, it is possible to interpret seismological recordings in terms of the source of the wind and understand the mechanisms of how wind can affect ground motion. By accounting temporal relations between wind speed and ground motion and considering the frequency domain that is affected, it is possible to draw conclusions on whether the local wind field or rather secondary effects such as induced water waves from the Dead Sea or the Mediterranean Sea have an impact on the seismological measurements.

During the investigation period from February, 2014 till January, 2015 multiple occasions took place when Station MET1 registered mean wind speed above $10\,\mathrm{m\,s^{-1}}$. Figure 6.1 shows the mean wind speed obtained at MET1 (top) and EBS3 (bottom) for the whole period. From the two winter seasons multiple periods of increased wind speed with different origin and properties, in the following referred to as events, are examined as case studies. The events are highlighted by dark green background in Figure 6.1: Events M1 and M2 in March, 2014, and Event J1 in January, 2015. They are selected on the basis of mean wind speed measurements at MET1 and by synoptic considerations. Note, that despite the small distance between MET1 and EBS3 of 15 km, the wind speed at EBS3 is considerably lower than at MET1, only rarely exceeding $5\,\mathrm{m\,s^{-1}}$, and with the exception of Event J1, periods of increased wind speed during the winter seasons at MET1 can not be identified at EBS3.
The following sections discuss each of the selected events by means of local wind speed at MET1 and EBS3 and synoptic processes on the basis of analysis from the numerical weather forecasting model GFS. The implications on seismological recordings are discussed for each event separately by analysing PSD and spectrograms for multiple stations. Analysis, used in this

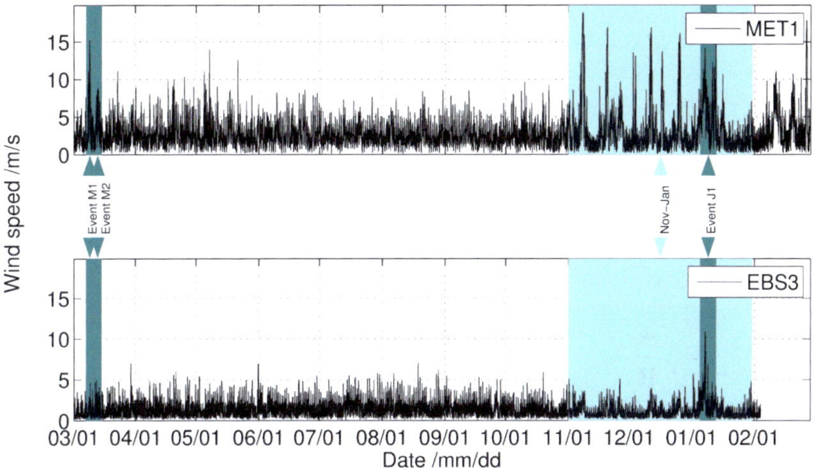

Figure 6.1: Measured mean wind speed during the measurement period from 01.03.2014 to 01.02.2015. Coloured periods mark investigation periods. Dark green periods mark winter storms investigated as case studies, light green markers highlight the long investigation period winter.

chapter, is calculated using methods that are specifically explained in Chapter 5.2. Basic processing of seismological data is described in Chapter 5.1, and unless otherwise specified, settings such as filters and time windows match given information.

6.1 Storm in January 2015: Event J1

In January, 2015, the storm with the largest temporal and spatial extent within the project period of this study took place in the Eastern Mediterranean. Due to the ongoing strong wind for a duration of more than two weeks over a large region it provides a suitable setting to investigate seismological records during the presence of strong wind.

6.1.1 Meteorological Situation

In the winter season of 2014 to 2015 several events with mean wind speed above 15 m s^{-1} occurred, and in large part, the mean wind speed at Station MET1 exceeded 10 m s^{-1} (Figure 6.1). Therefore, it is ambiguous to define the onset and the end of the period of one event. Here, the period between 06.01.2015 and 14.01.2015 is defined as Event J1, analysed for meteorological conditions and the seismological recordings in context.

The GFS model reveals the synoptic situation for the region of the Eastern Mediterranean. Figure 6.2 shows the GFS analysis for the first days of Event J1 from 06.01.15 to 08.01.2015, illustrating the 500 hPa geopotential, surface pressure, and the relative topography for the whole Mediterranean (a), the Eastern Mediterranean (b to d), and the wind field over the Eastern Mediterranean (e). The GFS analysis reveals a strong upper level trough over the Mediterranean Sea on 06.01.2015 (a). The ridge of the trough is located west of Israel and a low pressure system with a core pressure of 996 hPa at the surface moves towards northeast. As a consequence, a surface pressure gradient of approximately 5 hPa per 100 km evolves over the eastern Mediterranean Sea, Israel and Jordan. The wind turns its axis from southwest towards northwest while the low pressure area moves from 06.01.2015 to 08.01.2015 (b to d). Over the whole Mediterranean Sea increased surface wind speed of up to 25 m s^{-1} evolve from southwest. The most severe wind confines to the Mediterranean Sea. Over the land surfaces the wind weakens, with only slight increases of up to 10 m s^{-1} and directions from south and southwest. Confined to a resolution of 0.5°, the model is not able to resolve the small scale increased wind speed over Israel and Jordan, where local effects are controlled by the complex terrain of the Dead Sea valley (cf. Chapter 3.1.3). Though the model interprets comparatively low wind speed over the Dead Sea, it presents a storm in the Eastern Mediterranean that is likely to impair the local meteorological conditions at the Dead Sea.

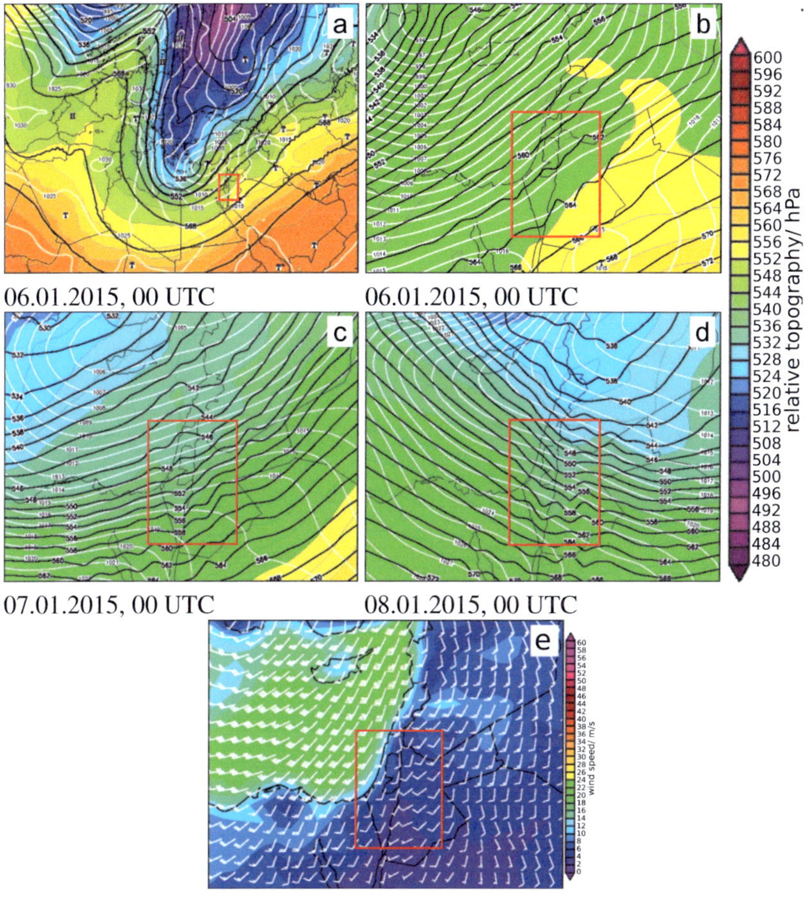

06.01.2015, 00 UTC | 06.01.2015, 00 UTC

07.01.2015, 00 UTC | 08.01.2015, 00 UTC

06.01.2015, 00 UTC

Figure 6.2: Synoptic situation during Event J1. **a-d:** synoptic scale evolution of 500 hPa geopotential (black) /*gpdm*, surface pressure (white) /*hPa* and relative topography 500-1000 hPa (coloured) /*gpdm*. **e:** wind speed and wind direction 10 m above ground. Colour scale in m s^{-1}, arrows in knot. Red frames encompass Dead Sea region. Data from GFS analysis, ©Wetter3.de

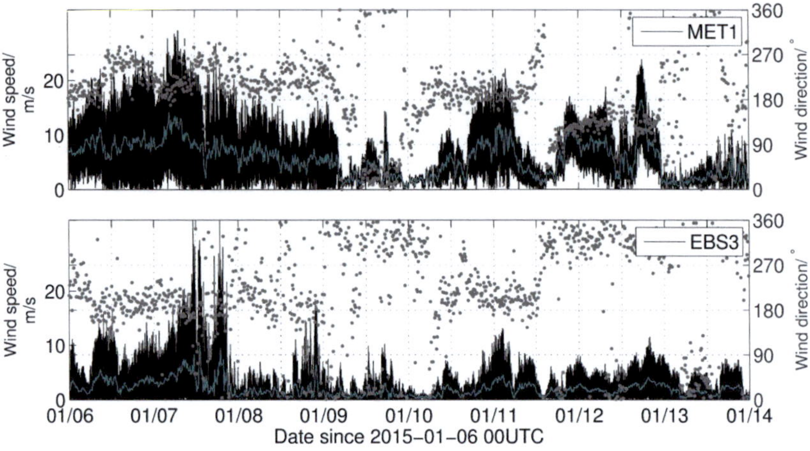

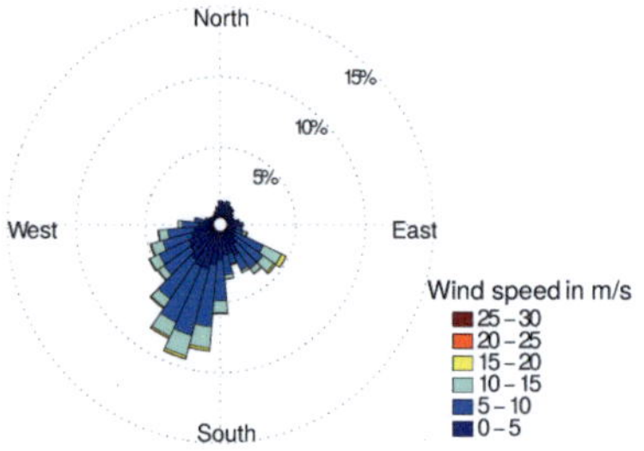

Figure 6.3: Top: Wind speed and wind direction at MET1 (top) and EBS3 (bottom) during Event J1 from 06.01.2015 to 14.01.2015. Wind speed in black, wind direction in grey, and 10-minute mean wind speed in green colour. **Bottom:** Wind rose from 06.01.2015 to 14.01.2015 at MET1.

The meteorological records from stations MET1 and EBS3 confirm the conclusions from the GFS model. Figure 6.3 shows wind speed and wind directions obtained at MET1 (top) and EBS3 (bottom) from 06.01.2015 to 14.01.2015. In addition to the high frequent wind speed data, the 10-minute average wind speed is given as green graph as well.

Throughout the defined period station MET1 observes high wind speed of up to $28\,\mathrm{m\,s^{-1}}$, alternating with short periods of low wind speed down to $5\,\mathrm{m\,s^{-1}}$. The time window from 06.01.2015 to 09.01.2015, the night of 11.01.2015, and the day of the 12.01.2015 in particular show high wind speed at MET1. The wind direction during J1 varies. However, strong winds from west and south dominate. Merely, on 12.01.2015 the strongest winds come from southeast.

At station EBS3 the record of the wind speed measurements is different to MET1. The maximum wind speed observed has more than 25 m s^{-1} on 07.01.2015 for a few hours. Only on 06.01.2015 and 07.01.2015 EBS3 observes the highest wind speed, most time above 10 m s^{-1}, however only half the speed than the wind speed of about 20 m s^{-1} at MET1 during the same period. The wind mainly comes from south and southwest directions, similar to the wind direction at MET1. During the period with lower wind speed from 08.01.2015 to 14.01.2015 the wind direction changes between northwest and south.

Although both stations MET1 and EBS3 show increased wind speed, the evolution of the local wind field differs and the overall wind speed is higher at MET1. But the wind comes from different directions. As the stations are located only 20 km off in horizontal direction, their sites are at different elevations and on opposite slopes of the Dead Sea Valley (Chapter 4.2). The atmosphere within the valley is often decoupled from the synoptic situation resulting in a deviating local wind field. Station EBS3 at the bottom of the valley experiences weaker influence by the synoptic situation than MET1 located 550 m above the valley floor. Due to the deep depression of the Dead Sea valley a uniform wind field similar to a plain site can not be expected.

The wind measurements at MET1 widely confirm the GFS analysis in magnitude and direction. Despite the low wind speed over Israel and Jordan in the model, the impact of the low pressure area can be clearly identified in the local measurements. It indicates that local effects due to the topography, not resolved by the model, influence the actual conditions.

On 11.01.2015 and 12.01.2015 the wind speed at MET1 is still high. However, the synoptic situation has changed. The upper level trough from Figure 6.2a still exists but has moved towards east with its axis located over the Dead Sea area now (Figure 6.4). The increased wind speed on 11.01.2015 and 12.01.2015 at MET1 and EBS3 can be linked to the surface pressure field north of Israel at the coast of the Mediterranean Sea. The wind measurements at MET1 accord with the GFS analysis, however, the model reveals that the high wind speed in the second period of Event J1 are caused by more local differences in the surface pressure field. The wind in the GFS analysis does not show strikingly increased values, neither over the Mediterranean Sea, nor over the land surfaces.

6.1.2 Wind Induced Seismic Signals during Event J1

The presence of air flow along the earth's surface can excite the ground similar as tectonic ruptures or anthropogenic influences can do. Therefore, seismological stations register signals that may correlate with the presence of wind. To investigate potential correlations of wind and the seismological records time series of both datasets are collocated and compared. The seismological sensors record ground motion velocity as a function of time. However, to investigate not only the signal amplitudes with respect to time, but also the frequency content of signals to gain insight into the signal properties, spectrograms of the ground motion velocity are calculated and analysed (Chapter 5.2).

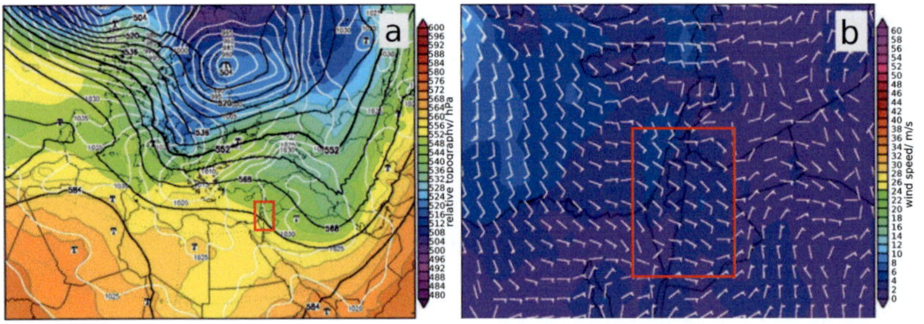

12.01.2015, 00 UTC   12.01.2015, 00 UTC

Figure 6.4: Synoptic situation during second part of Event J1. **a:** synoptic scale evolution of 500 hPa geopotential (black) /gpdm, surface pressure (white) /hPa and relative topography 500-1000 hPa (coloured) /gpdm. **b:** wind speed and wind direction 10 m above ground. Colour scale in m s^{-1}, wind arrows in knot. Red frames encompass Dead Sea region. Data from GFS analysis, ©Wetter3.de

In this section, seismological spectrogram properties of selected seismological stations are investigated and compared to data of wind records from the meteorological station MET1. Specific methods and characteristics to obtain the spectrograms of ground motion velocity and further methods are described in detail in Chapters 5.2 and 5.2.3. This section omits the discussion of recordings from horizontal seismological records focussing only on the vertical components to be independent from the influence of directivity of waves. Comparisons of seismological records in different components are subject to Section 6.3 and Chapter 7. Results and conclusions also apply to data from stations not discussed in detail here.

PSD Level at all Stations during J1

As to gain a first impression on the seismological records of all stations during a storm, such as Event J1, one may consider the PSD for a selected period. The PSD resolves the power level of the seismological measurements with respect to the frequency. For frequencies above 1 Hz one can expect to have an influence of signals from local sources, such as traffic, other anthropogenic noise, and regional earthquakes, but also the influence of wind on ground movement in the area. Apart from seismic signals the level of the PSD can vary individually for seismological stations, depending on instrument type, station type (permanent or temporary), site, or region. It is inadequate to interpret from one sample period only the shape and the level of the PSD data in terms of signal sources. By comparing selected datasets with opposed wind conditions but similar conditions in terms of anthropogenic noise it is possible to draw conclusions on the influence of wind on the seismic recordings.

Two datasets of 6 hours length each are selected: a time window between 18 UTC and 00 UTC on 01.03.2014, when MET1 provides a wind speed below $3\,\mathrm{m\,s^{-1}}$, and a second period on 06.01.2015 from the same time of the day when MET1 provides the maximum wind speed during Event J1. During the chosen night time one can expect the influence of anthropogenic noise on the data to be minimal.

Figure 6.5 shows the PSD for all seismological stations on the Z-component in frequencies between 0.1 Hz and 100 Hz for both periods. On 01.03.2014 the PSD varies between less than 0 dB at minimum in frequencies above 20 Hz and up to 45 dB in frequencies below 1 Hz. In frequencies above 20 Hz the data have the lowest PSD values not exceeding 20 dB. Between 1 Hz and 20 Hz the PSD level is higher, ranging between 5 dB and 30 dB. Below, the trends of the PSD graphs rise up to 45 dB before declining. The PSD levels of data from stations that are equipped with short period sensors (Chapter 4.1) drop below 0 dB below 0.2 Hz due to filtering. Altogether, the data quality of all stations is well within the expected range for temporary seismological stations.

In comparison to 01.03.2014, the PSD on 06.01.2015 are on an increased level. The PSD levels are raised by at least 10 dB and up to 35 dB, especially in the frequencies above 20 Hz, but also the absolute maximum at 0.2 Hz has increased to more than 60 dB for the broad band stations. Above 1 Hz the PSD levels drift at all stations, below the graphs are more uniform, similar to the PSD graphs in the left panel, but raised by about 20 dB.

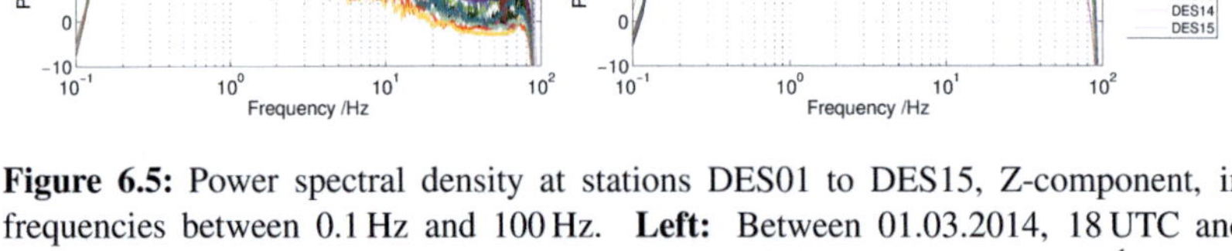

Figure 6.5: Power spectral density at stations DES01 to DES15, Z-component, in frequencies between 0.1 Hz and 100 Hz. **Left:** Between 01.03.2014, 18 UTC and 02.03.2014, 00 UTC while mean wind speed at MET1 does not exceed 3 m s^{-1}. **Right:** Between 06.01.2015, 18 UTC and 07.01.2015, 00 UTC with mean wind speed at MET1 exceeding 10 m s^{-1}. Data are smoothed sectionwise to allow better visibility.

Within periods of 6 hours short signals of seconds or even a few minutes length such as local or regional earthquakes would not have a major impact on the PSD level of the ground motion velocity. By examining the PSD levels at all seismological stations for a period of six hours, it is possible to relate changes of the PSD levels to sources with long durations, such as it is the case for the local wind. The comparison of data from the calm wind situation on 01.03.2014 with data from 06.01.2015 during Event J1, strongly indicates that the striking increase of the PSD levels is caused by the wind with high speed during Event J1.

Seismological Spectrograms during J1

Spectrograms can confirm that wind is the most plausible cause for high power densities of the seismological records. To clearly identify signals related to a long lasting meteorological situation, the observation of an even longer period can be helpful. Figure 6.6 shows spectrograms of data from station DES15, the closest seismological station to the meteorological station MET1, for January, 2015. The spectrograms display the PSD of ground motion velocity with respect to the frequency and to time. The spectrogram in the top panel of Figure 6.6 comprises the frequency domain from 1 Hz to 90 Hz, the spectrogram below zooms in to frequencies between 0.1 Hz and 1 Hz. For comparison, the bottom panel displays the mean wind speed measured at MET1 with respect to the same time axis. Other than in Figure 6.5 the relative, logarithmic scale of the PSD is given not with respect to $1\,\mathrm{nm}^2\,\mathrm{s}^{-2}\,\mathrm{Hz}^{-1}$ but relative to the signal maximum of $3.34\mathrm{E}{+}06\,\mathrm{nm}^2\,\mathrm{s}^{-2}\,\mathrm{Hz}^{-1}$ at the specific station in the present period.

The spectrograms reveal strong signals in both frequency domains that correlate temporally with the high wind speed from 02.01.2015 until 14.01.2015 at MET1. The strongest impact can be found in frequencies below 1 Hz where one strong, long signal can be found between 0.1 Hz and 1 Hz lasting from 02.01.2015 to 18.01.2015 matching the period of increased wind speed above

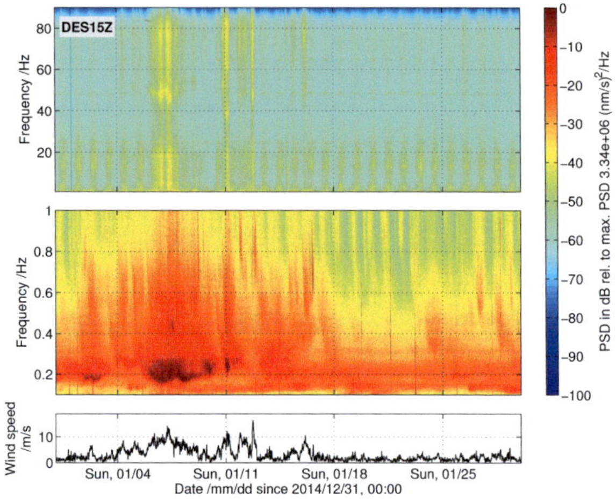

Figure 6.6: Spectrograms of station DES15, Z-component, between 10 Hz to 90 Hz (top), 0.1 Hz to 1 Hz (middle), and mean wind speed at meteorological station MET1 (bottom). All graphs with respect to the same time axis. PSD in dB relative to maximum PSD.

$5\,\mathrm{m\,s^{-1}}$. The range of the frequency domain with high PSD extends with increasing wind speed. Furthermore, the magnitude of the PSD level increases with increasing wind speed. The highest values of PSD can be identified in a narrow frequency band around 0.2 Hz which is most striking for the period from 06.01.2015 to 12.01.2015 where the mean wind speed has a persistently high level.

In frequencies above 1 Hz the signals related to the wind speed regime are less strong, however clearly apparent. The signals comprise the whole frequency domain up to 90 Hz but only shorter periods of time from 06.01.2015 to 08.01.2015, 10.01.2015 to 13.01.2015, and 16.01.2015 coinciding with high peaks of wind speed above 10 Hz. The intensity of the PSD signals in the high frequencies is about 30 dB lower than the signal maxima in the low frequencies. However, the overall PSD level in the absence of strong wind is generally higher in the low frequencies. The intensity of the impact of wind can therefore be estimated to be comparable in frequencies above and below 1 Hz.

In addition to the signals in the frequencies above 1 Hz, signals of approximately ten hours length occur diurnally during day times. They are weaker than the wind related signals and due to their periodical recurrence they can be assigned to anthropogenic influence.

In this section, signals at stations DES01, 03, 11, 13, 15, and GHAJ, each for the Z component, are under investigation. Regarding two weeks of spectrograms from 02.01.2015 to 14.01.2015, the impact of Event J1 on ground motion velocity can be investigated in more detail. Figure 6.7 comprises spectrograms of broad band stations DES01, DES13, DES15, and short period stations DES03 and DES11, as well as the permanent broad band station GHAJ for each one frequency domain above and below 1 Hz. Note, that the lowest frequency displayed in the spectrograms of the short period stations is 0.2 Hz instead of 0.1 Hz at the broad band stations. The highest frequency to calculate spectrograms for is determined by the sampling rate of the seismological records (Chapter 4.1), therefore the spectral data of station GHAJ is bound by an upper limit of 50 Hz, compared to 100 Hz at the DES stations.
In frequencies above 1 Hz temporal correlations of high PSD values and high wind speed can be clearly identified for all stations but DES01. At DES01 increased signals related to the highest wind speed peaks on 07.01.2015 and 11.01.2015 to 13.01.2015 can be identified, too, but the signal intensities are less dominant. In frequencies above 1 Hz the signals at each individual station vary in duration, frequency domain, shape, and intensity. Nevertheless, all signals correlate with peaks of the highest wind speed, most distinct where the wind speed increases and decreases. Some of the stations have the highest signal impact in specific frequency bands, such as for example station DES01 between 60 Hz and 80 Hz, station DES13 between 40 Hz and 85 Hz, or station GHAJ in multiple narrow streaks in the whole domain above 1 Hz. The signals in frequencies above 1 Hz clearly temporally correlate with the local wind speed without any significant time delays.

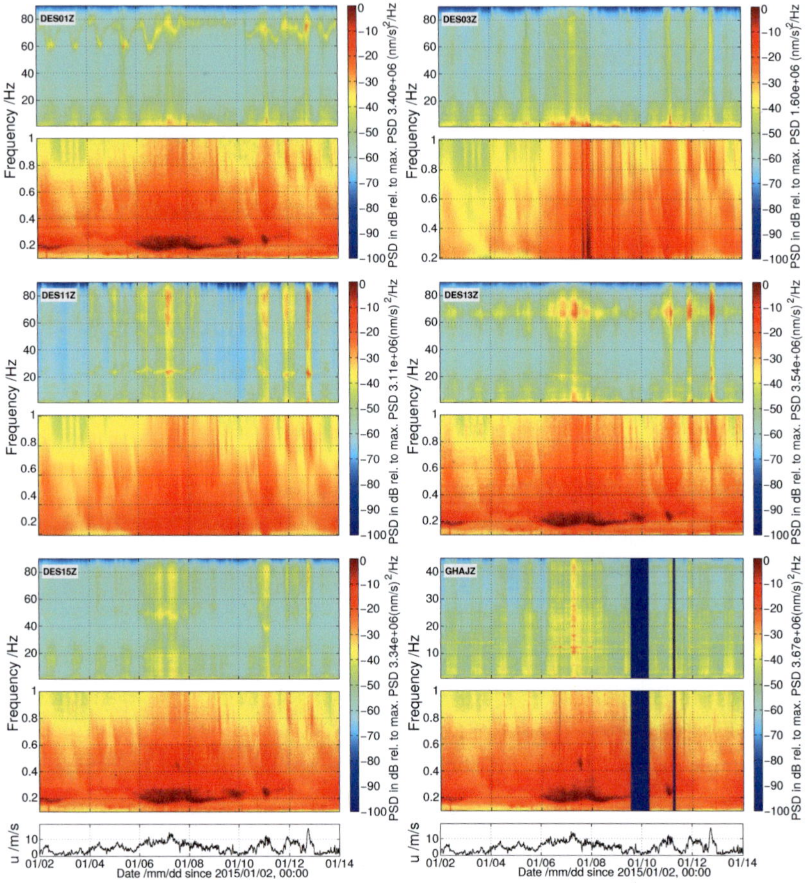

Figure 6.7: Spectrograms of stations DES01, 03, 11, 13, 15, GHAJ Z-component, during Event J1. For each seismological station, two separate spectrograms for frequencies below and above 1 Hz are given. PSD is displayed as relative scales depending on the maximum at each separate record. Wind speed record *u* from station MET1. Dark blue bands in spectrograms of GHAJ mark data gaps.

Observing the spectrograms below 1 Hz provides other conclusions. Like found for station DES15 in Figure 6.6 strong signals can be found in frequencies below 1 Hz at all stations. Taking into account the different lower frequency limits in Figure 6.7 at each of the stations fairly similar, long, and strong signals can be observed for the whole period of Event J1. When regarding broad band and short period stations separately, the signals are nearly identical. The signal maxima are located in a narrow band between less than 0.2 Hz and about 0.3 Hz, most striking for the period between 05.01.2015 and 12.01.2015. Reckoning that the highest local wind speed peak is measured between 13.01.2015 and 14.01.2015, dissociated from highest PSD signal period and that during the long period of the highest PSD from 08.01.2015 to 11.01.2015 the average wind speed drops below $10\,\mathrm{m\,s^{-1}}$, the strong signal maximum is not directly correlated with the local wind speed unlike the signals visible in the high frequencies. The fact that the correlation of the signals in the lower frequency domains is high, furthermore indicates that at least a significant fraction of the signal is not related to the local wind but to a source in farer distance. The congruent signal at station GHAJ with 50 km offset to the DES stations supports this hypothesis.

The spectrograms during Event J1 confirm the impact of wind on seismological records. It is even possible to distinguish between the impact on different frequency domains and to allow first conclusions on the signal source of the excitation of ground motion velocity. Seismological records are affected by high wind speed in the broad frequency domain at least between 0.1 Hz and 100 Hz investigated here. Above 1 Hz the signals have individual properties at each seismological station and correlate without visible time lag with the course of the local wind speed obtained from MET1. In lower frequencies below 1 Hz the impact has higher signal magnitudes for Event J1, coming along with uniform signal properties for each station, independent of local effects. The spectrograms therefore imply that the signal source of the signals below 1 Hz is strong and distant. The GFS analysis in Figure 6.4 confirms that during

Event J1 strong winds exist in the Eastern Mediterranean, especially above the Mediterranean Sea which can be strongly considered as signal source in the frequency band below 1 Hz. The following sections take up the questions that arise with the signal correlations of the seismological spectrograms and the wind speed data, such as the identification of signal sources (Section 6.3), and the signal intensities in dissociation to common tectonic signals (Section 6.5).

6.2 Wind Speed in March 2014: Events M1 and M2

March 2014 is the first month where the meteorological stations MET1, EBS3 and all of the seismological stations operated simultaneously providing a complete dataset. It is the last month of the winter season of 2013/2014 and at the same time the first period within the measurements when the meteorological station MET1 measured high mean wind speed of up to $15\,\mathrm{m\,s^{-1}}$. Despite similarly strong wind as during Event J1, the synoptic situation in March is completely different and therefore worth investigating.

6.2.1 Meteorological Situation

Between 07.03.2014 and 14.03.2014 the wind measurements at MET1 show two consecutive events with increased wind speed (cf. Figure 6.1). During the first period from 06.03.2014 to 08.03.2014 the mean wind speed at MET1 reached maximum values of $15\,\mathrm{m\,s^{-1}}$, in the following this period is referred to as Event M1. The period is followed by another incident, where the mean wind speed at MET1 was increased for a period of two days from 12.03.2014 to 14.03.2014, referred to as Event M2, but did not exceed $10\,\mathrm{m\,s^{-1}}$.
To link the locally obtained wind speed with its driving force it is essential to consider the synoptic situation on a larger scale by examining GFS analysis data for the Mediterranean and the Middle East. Figure 6.8a-e illustrates the GFS analysis for the period from 08.03.2014, 00UTC to 09.03.2014, 00UTC. On 08.03.2014, a trough over the Mediterranean Sea with an axis from Libya towards Eastern Europe determines the meteorological conditions in the Dead

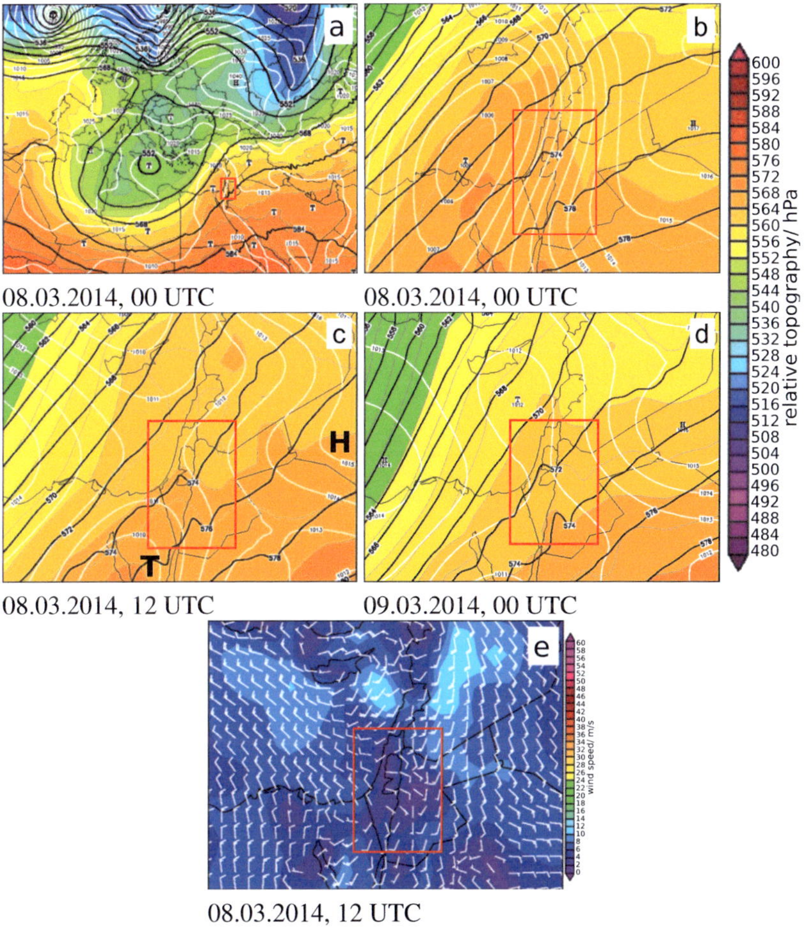

Figure 6.8: Synoptic situation during Event M1. **a-d:** synoptic scale evolution of 500 hPa geopotential (black) $/gpdm$, surface pressure (white) $/hPa$ and relative topography 500-1000 hPa (coloured) $/gpdm$. **e:** wind speed and wind direction 10 m above ground. Colour scale in $m\,s^{-1}$. Arrows in knot. Red frames encompass Dead Sea region. Data from GFS analysis, ©Wetter3.de

Sea area. In front of the trough near the surface a low pressure area (Panels a and b) develops, moving from Egypt towards Turkey off the coast of Israel causing a pressure gradient from the Mediterranean Sea towards Jordan. The pressure gradient is strongest on 08.03.2014, 00 UTC (Panel b) with roughly 2 hPa per 100 km causing wind of about $10\,\mathrm{m\,s^{-1}}$ from southeast over the Dead Sea (Panel e). According to the model, on 08.03.2014, 12 UTC and 09.03.2014, 00 UTC the pressure gradient is less pronounced and the wind over the Dead Sea area is weaker.

According to Figure 6.8 the wind at the Dead Sea is strongest on 08.03.2014, 00 UTC. The measurements at MET1 dissent, obtaining the maximum wind speed between 08.03.2014, 18 UTC and 09.03.2014, 03 UTC when the GFS analysis reveals a less pronounced pressure field. The GFS model with a resolution of 0.5° may be unable to resolve the local processes, such as slope winds, driven by the complex topography (cf. Chapter 3.1.3).

The general weather situation determined by the trough in the GFS analysis results in increased wind in Jordan and Israel. However, the data do not reveal a major storm caused by a distinct low pressure system. These observations lead to the assumption that the strong wind at MET1 might be a local manifestation of high wind speed within overall turbulent atmospheric conditions and therefore one can not expect strong wind on large-scales, having, for example, an impact on the Mediterranean Sea like it is the case during Event J1 (cf. Section 6.1.1). Instead, the local wind field may be dominated by processes on small scales, such as the topography, enforcing wind channelling inside the Dead Sea valley, slope winds along the valley flanks, and potential decoupling of the small-scale systems from the atmosphere.

On 11.03.2014 the trough above the Mediterranean Sea still exists (Figure 6.9a, cf. Figure 6.8), however, the axes of the trough moved east and is now located over the Dead Sea. The situation evolves a gradient of surface pressure of about 2 hPa per 100 km from north to south across the Dead Sea and the Israel coast of the Mediterranean Sea (Panel b). Driven by the pressure gradient wind speed of up to $16\,\mathrm{m\,s^{-1}}$ develop over the Mediterranean Sea and east of the

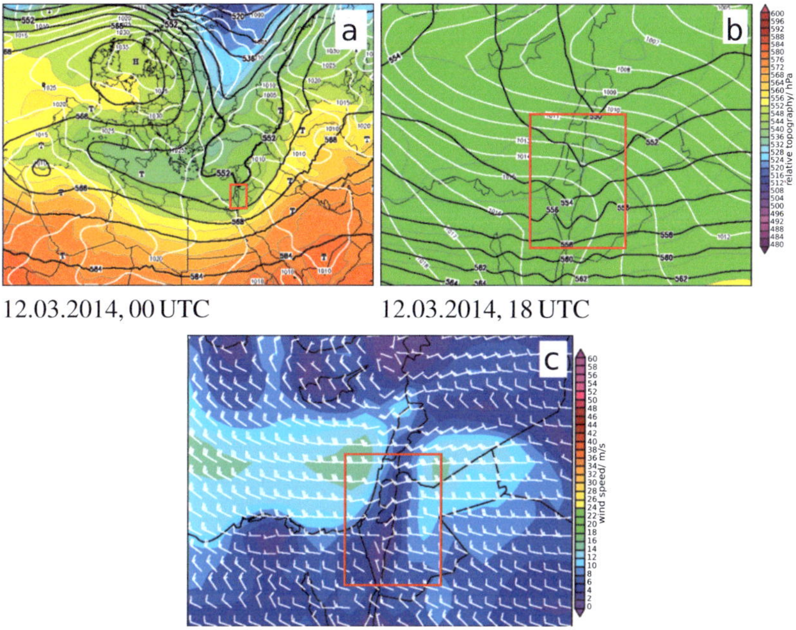

12.03.2014, 00 UTC

12.03.2014, 18 UTC

12.03.2014, 18 UTC

Figure 6.9: Synoptic situation during Event M2. **a-b:** synoptic scale evolution of 500 hPa geopotential (black) /gpdm, surface pressure (white) /hPa and relative topography 500-1000 hPa (coloured) /gpdm. **c:** wind speed and wind direction 10 m above ground. Colour scale in m s^{-1}. Arrows in knot. Red frames encompass Dead Sea region. Data from GFS analysis, ©Wetter3.de

Dead Sea in the high plains of Jordan and Syria (Panel c). Across the Dead Sea valley the wind is weaker and comes from southwest.

Altogether, like for Event M1, the GFS model reveals increased wind speed for the Dead Sea region, but no major storm. Other than for M1, the wind speed in the model is higher for M2, however, the maximum wind speed exists over the Mediterranean Sea and east of the Dead Sea with a local minimum over the Dead Sea where no increased wind speed is given.

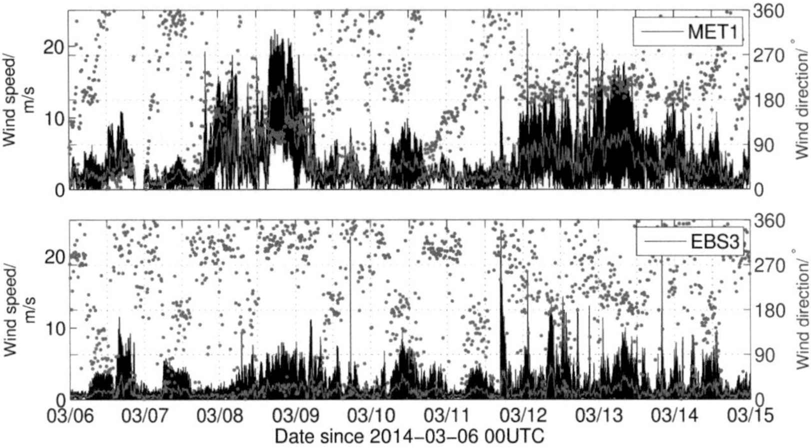

Figure 6.10: Wind speed and wind direction at MET1 (top) and EBS3 (bottom) from 06.03.2014 to 15.03.2014. Wind speed in black, wind direction in grey, and 10 minute mean wind speed in green colour.

To assess the nature of the local wind during Events M1 and M2, records of both stations MET1 and EBS3 should be considered. Figure 6.10 shows time series of wind speed and wind direction at MET1 and EBS3 from 06.03.2014 to 15.03.2014 covering Events M1 and M2. In contrast to the mean wind speed in Figure 6.1, the black graphs illustrate recordings of turbulent wind data (20 Hz). For comparison, the green graphs in both panels represent the 10 minute mean wind speed.

At station MET1 the two periods with maximum mean wind speed of up to $15\,\mathrm{m\,s^{-1}}$, and $8\,\mathrm{m\,s^{-1}}$ respectively, that make Events M1 and M2, can be clearly identified. During Event M1 the wind speed at MET1 increases from less than $5\,\mathrm{m\,s^{-1}}$ to more than $20\,\mathrm{m\,s^{-1}}$ for the period from 07.03.2014, 18 UTC to 09.03.2014, 06 UTC. MET1 observes the highest wind speed with the main wind direction between 100° and 130° in the evening of 08.03.2014.

Event M2 is less striking in the wind speed record from MET1 with maximum wind speed of up to $20\,\mathrm{m\,s^{-1}}$, but mean wind speed of less than $10\,\mathrm{m\,s^{-1}}$

between 12.03.2014 and 14.03.2014. The main wind direction in this period ranges between 190° and 270°. Altogether, during Event M1 MET1 measures higher wind speed than during Event M2. It becomes particularly obvious when observing the mean wind speed instead of the turbulence. The 10 minute mean during M2 does not exceed $8\,\mathrm{m\,s^{-1}}$ which is only half of the peak mean wind speed during M1.
During the entire period, the wind speed at MET1 is apparently higher than at EBS3, this is especially the case for the strongest wind during the events. Despite the inferior site at EBS3 between natural vegetation, expected to decelerate the wind (Chapter 4.2), one would expect an attenuated, but similar wind field compared to MET1. With this assumption it is likely that not the quality of the site of EBS3 is the cause for the noticeable differences in the wind fields but rather the actual wind field at the location of EBS3 differs. EBS3 is located about 500 m deeper than MET1 and on the opposite site of the valley (Figure 4.15). So rather the data in Figure 6.10 indicate that the high wind speed, observed during the severe wind events M1 and M2 at MET1, does not reach the bottom of the valley at the location of EBS3. Between the altitudes of EBS3 and MET1 one can assume a boundary in the atmosphere where the part below is decoupled from larger or synoptic processes. The differences in the local wind field at MET1 and EBS3 can be observed for the whole project period, as it becomes clear in Figure 6.1, also discussed for Event J1 in January in Section 6.1.1 (Metzger et al., 2016).

In summary, Event M1 stands out due to high wind speed recorded at station MET1, the maximum mean wind speed exceeding $15\,\mathrm{m\,s^{-1}}$. While the mean wind speed during M2 does not exceed $10\,\mathrm{m\,s^{-1}}$ an increase of wind speed is difficult to recognize. On synoptic scale, both periods can not be explained by major storms. For the area of the Dead Sea the model is not able to resolve increased wind speed at all. Solely for Event M2, where the wind speed increase, measured at MET1, is more unremarkable than during M1, the model reveals increased wind speed over the Mediterranean Sea and in West Jordan. The fact, that other than the measurements of high wind speed obtained at MET1, the model does not show increased wind speed during M1, indicates that M1 is a very local event that can not be resolved by the GFS model. Compared to Event J1 in January 2015 the wind speed during M1 is of an equivalent level at MET1, however the GFS model reveals that unlike J1 Event M1 is rather a mesoscale than a synoptic event.

6.2.2 Wind Induced Signals during Events M1 and M2

PSD Level at all Stations During M1 and M2

Like for Event J1, a first impression on the ground motion velocity during the observational period can be achieved by examining the PSD for selected periods. Periods of interest are time windows with preferably the highest wind speed during the events coinciding with night times where the influence of anthropogenic noise is lowest.

Figure 6.11 shows the PSD of the ground motion velocity in the Z-components at all DES stations for periods of six hours within Events M1 and M2 for frequencies between 0.1 Hz and 100 Hz. Two time windows are selected: the left panel shows PSD levels beginning 08.03.2014, 18 UTC while the wind speed at MET1 reaches its maximum during M1. The right panel illustrates the corresponding data starting 13.03.2014, 00 UTC when the wind speed is highest during M2. Note, that both periods comprise time windows at night to minimize the potential anthropogenic influence on the seismological records, but due to the short durations of high wind speed periods the time windows do not cover the same times of day.

Compared to the calm wind situation on 01.03.2014, when the noise level at the seismological stations varies between less than 0 dB and 45 dB at maximum (Figure 6.5, left panel), the level of the PSD is increased at all stations during both Events M1 and M2. The PSD levels during Event M1 range between approximately 15 dB and 60 dB at maximum, whereas the divergence between the different stations is biggest for frequencies above 10 Hz. Below frequencies

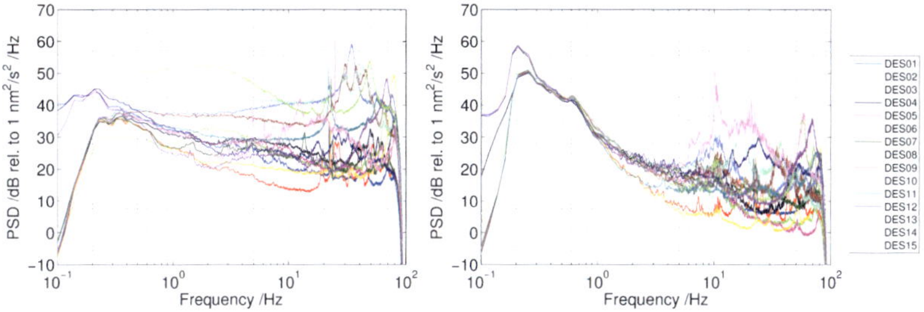

Figure 6.11: Power spectral density at stations DES01 to DES15, Z-component, in frequencies between 0.1 Hz and 100 Hz. **Left:** Between 08.03.2014, 18 UTC and 09.03.2014, 00 UTC during Event M1 while mean wind speed at MET1 reaches up to 15 m s^{-1}. **Right:** Between 13.03.2014, 00 UTC and 13.03.2014, 06 UTC with mean wind speed at MET1 reaching 10 m s^{-1}. Data are smoothed to allow better visibility of all graphs.

of 0.2 Hz the PSD levels of the short period instruments decay and the PSD of the broad band instruments reach a maximum of 45 dB. An exception is data from station DES13 showing an abnormal course: the magnitude of the PSD increases with decaying frequencies below 1 Hz till reaching a level 60 dB at 0.1 Hz. In comparison to the period during a calm wind situation (Figure 6.5) the PSD levels are increased by about 15 dB in frequencies above 10 Hz, but in frequencies below the graphs run similar courses on the same levels.

On 13.03.2014 the PSD levels have different shapes. Compared to Event M1 the trends of the graphs perform diametrical. Above 10 Hz the PSD levels are only marginally increased compared to the calm wind situation. But below 10 Hz all graphs increase to maxima of 50 dB, respectively 60 dB at 0.2 Hz resulting in an amplification of 15 dB compared to Event M1 and the calm situation. The peaks at 0.2 Hz almost reach the peak PSD level during Event J1 in January.

As presumed by the examinations of the events the PSD estimates indicate that the Events M1 and M2 represent different meteorological situations. While during Event M1 an increased level of PSD is observed primarily in frequencies above 1 Hz, for M2 it is the other way round with increased values of PSD for frequencies only below 2 Hz.

Seismological spectrograms during M1 and M2

As Events M1 and M2 follow each other in succession and have unique properties that are worth comparing, spectrograms in this section cover periods from 07.03.2014 to 16.03.2014 that comprise both events. To achieve a representative overview on the situation at the seismological stations, like for Event J1 (cf. Section 6.1.2) multiple stations of different type are selected: broad band stations DES01, 14, 15, short period stations DES06 and DES10, and permanent station GHAJ. Figure 6.12 displays each one spectrogram for frequency domains below 1 Hz and above 1 Hz for all stations.

In all spectrograms illustrated in Figure 6.12 the impact of both events is clearly visible. In frequencies above 1 Hz Event M1 on 06.03.2014 to 08.03.2014 is dominant, Event M2 is weaker, but can be identified best in data from stations DES06, DES10, and DES15. At station DES15, which is closest to the meteorological station MET1, the temporal correlation between increased power spectral densities above 1 Hz and the course of the wind speed is most striking. As observed for Event J1 in Section 6.1.2, aside from visible wind related signals at all stations, the spectrograms in the domain above 1 Hz have very individual appearances. The spectrogram of station DES14 in the high frequency domain makes an exception. It is dominated by strong diurnal signals that can be assigned to anthropogenic noise due the close vicinity of the Dead Sea highway (cf. map in Figure 4.2) superposing any other potential weaker signals in frequencies between 1 Hz and 100 Hz.

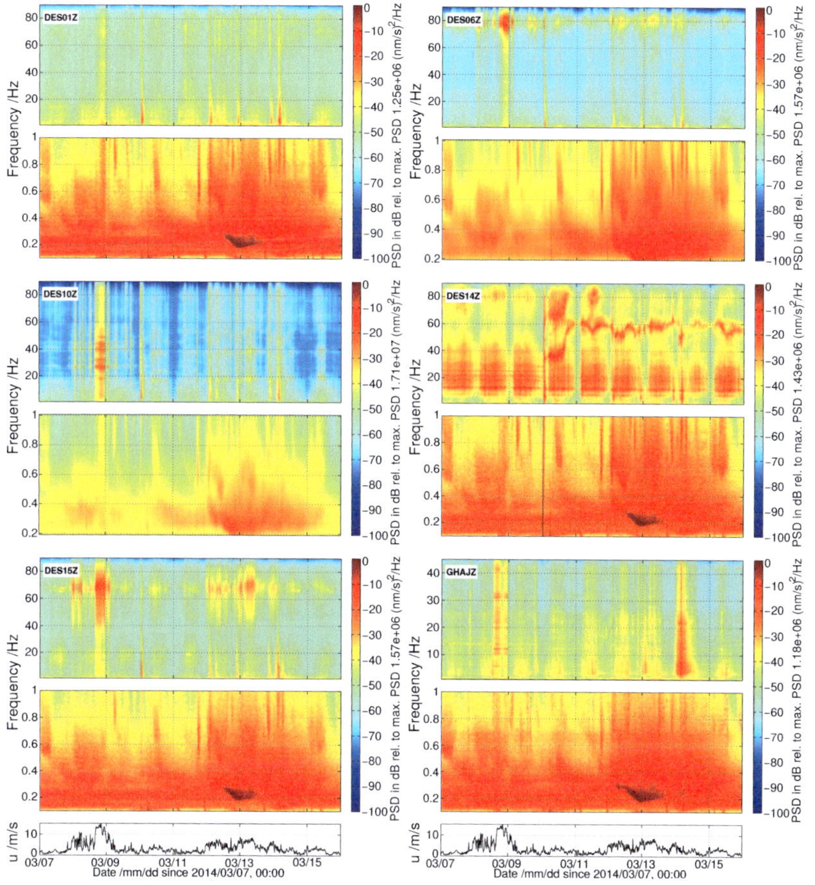

Figure 6.12: Spectrograms of stations DES01, 06, 10, 14 15, GHAJ, Z-component, during Events M1 and M2. For each seismological station, two separate spectrograms for frequencies below and above 1 Hz are displayed. PSD is displayed as relative scales depending on the maximum at each separate record. Wind speed records u are from station MET1.

In the frequency domain below 1 Hz the impact of M1 is barely visible, however, signals related to M2 from 12.03.2014 to 16.03.2014 are very dominant and contain the maximum power within the ground motion velocity during the period of the records at each station. The signals at each station are in strong accordance and have distinct maxima around 0.2 Hz. Compared to the low frequent signals during Event J1 the signals are similar. It should be emphasized that despite the high noise level at station DES14, where no wind-induced signals can be identified in frequencies above 1 Hz, the signal in the low frequency domain is as strong as at the other stations and the spectrogram appears identical compared to the other broad band stations. It confirms, that the influence of anthropogenic noise is mainly restricted to frequencies above 1 Hz.

Conclusions on M1 and M2

By examining both Events M1 and M2 in meteorological and seismological records, it is possible to draw conclusions on the source of wind related signals in the seismic records. Both events are small systems with moderate strength, hence the GFS model is not able to resolve Event M1 or to show remarkably high wind speed for Event M2. During M1 we obtain high local mean wind speed of up to $15\,\mathrm{m\,s^{-1}}$ causing strong signals in the ground motion velocity above 1 Hz that reflect directly the evolution of the wind speed. The fact that the signals obtained at station DES15, closest to the wind measurements, have the best correlation, indicates that the signals must be caused by the local wind field that is measured at MET1. The data also suggest that the signals in the same frequency domain at all stations indeed correlate instantly with wind speed changes, however, having individual shapes and intensities in various different frequency bands, specific for each station. The different shapes of the signals at all stations not only indicate that the source of these signal must be the local wind field, affecting near ground or obstacles, such as rocks or the edge of the valley, they also suggest that the ground around each station has unique properties such as specific eigenfrequencies and site effects.

The signals in the frequencies below 1 Hz in both events also appear simultaneously with locally increased wind speed. For both events, however the temporal correlation is more diffuse than in the high frequency range, single bursts or drops of the wind speed can not be identified in the spectrograms, only the whole time intervals of increased wind speed. The difference in both small events is not only the different local wind speed and resulting signals of different intensities in the high frequencies. The signals in the low frequencies reveal information on the excitation of seismological signals at farer distances and, therefore, do not only reflect the local situation. Despite lower wind speed at MET1 during Event M2 strong signals in the low frequencies can be assigned to the event.
Considering that the frequency domain with the maximum energies is around 0.2 Hz it is possible to assign the sharp maximum to ocean microseisms (cf. Section 6.3.2) from the Mediterranean Sea that are increased compared to situations without strong winds. Permanent station GHAJ is located 50 km south of the DES stations and records an almost identical signal. This coincidence implies that the signal source is not a local source. The seismological spectrograms confirm the GFS model and indicate that Event M2 is not only an event with increased wind speed, but a minor, regional storm with increased winds over a wider area covering at least the Eastern Mediterranean Sea. The diffuse but strong excitation in the spectrograms in the low frequencies during M2, surrounding the distinct maximum around 0.2 Hz, covers a broader frequency band up to 1 Hz, going along with the increased wind speed.

The conclusions from Events M1 and M2 confirm the observations from Event J1. From the comparison of seismological spectrogram data with wind speed records and regional GFS analysis questions arise such as to distinguish between actual signal sources in different frequency domains of seismological records or to quantify the correlation of local wind speed and seismological records. Another interesting point is the classification of wind-induced signals in terms of frequency domain and intensity in comparison to tectonic events.

The following Sections 6.3 and 6.5 discuss these questions based on the properties of the spectrograms during Events M1, M2, and J1.

6.3 Localisation and Identification of Signald Induced by Wind

Seismological data quality highly depends on the sites, the local, and the regional conditions of the area where the seismometer is deployed. The influence of wind induced stresses on the ground strongly depends on the point of contact where the force acts. Whereas we can approximate tectonic ruptures in adequately long distances, the far-field, as point sources, wind acts on large areas far away from the stations as well as on areas enclosing the station sites and therefore can not be treated and located like earthquake sources.

Figure 6.13 illustrates a schematic conception of how wind can affect the ground motion recorded by a seismological station. Wind, indicated as black arrows, acts on the earth's surface near the seismological station (red triangle) and at farer distances. The wind may hit the ground directly without any obstacles, or it can act on obstacles, such as trees and buildings, where not only shear stress, but also compressive stress affect the surfaces. The oscillations couple into the ground where seismic waves, indicated as red curved arrows, spread out. Strong amplifiers for wind acting on land surfaces are water bodies, such as the Dead Sea and the Mediterranean Sea. Wind causes shear stress above the sea that agitate ocean waves, propagating to the coasts and interfering. They transmit strong forces into the ground that can be recorded by seismological stations even in distances of hundreds of kilometres as narrow band, and low frequent signals. Since signals from water bodies and local ground movements have different properties at recording seismological stations, it is possible to distinguish the sources. Like waves from tectonic ruptures seismic waves induced by other sources are subject to geometrical spreading and attenuation. Depending on the propagation path, along the surface, or as a body wave in-

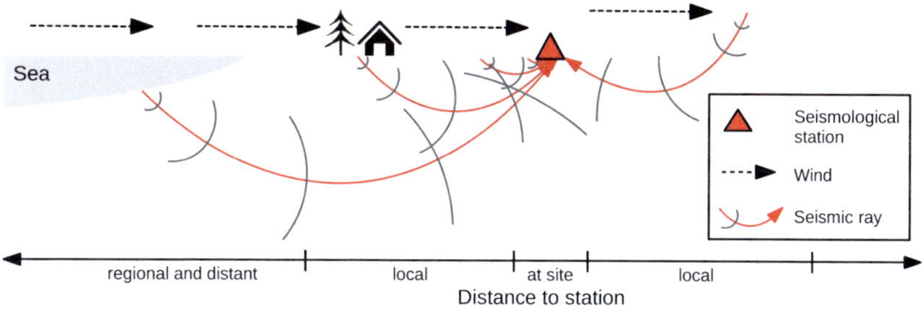

Figure 6.13: Excitation of wind-related seismological signals with respect to distance to the seismological station. The characteristics of wind-related signals depend on the raypath of the seismic waves and on potential obstacles, such as trees, buildings, and water bodies meeting the wind and acting as possible amplifiers. Red arrows indicate the seismic signals as rays. The perpendicular, bent grey lines indicate the course of the corresponding wave fronts.

side the earth, the wave arriving at the recording station is dissipated with a frequency-dependent loss of amplitude. Therefore, signals from far sources can be distinguished from signals from nearby sources by their lower amplitudes and the lower frequency content. As a result, based on examinations of signal amplitudes and the frequency domain, it is possible to estimate the distances between station and signal source. However, it is not possible to locate clearly separable contact points of wind-related signals as sources because the signal at the recording station comprises a superposition of the whole wave field.

Hereinafter, distances between signal sources and the seismological stations are divided into separable classes and defined as follows:

- Site: source in immediate vicinity of the station. No measurable effect at other stations
- Local: signal sources within hundreds of meters to up to a few kilometres around the array of the seismological stations

- Regional and distant: sources from several kilometres away to hundreds of kilometres away

The following sections examine the sources of wind related signals in seismological records and identify the influence of wind on large water bodies, such as the Mediterranean Sea.

6.3.1 Local Sources and Site Effects

The spectrograms of the seismological stations in Sections 6.1.1 and 6.2 are separately analysed in the frequency domains above and below 1 Hz. The examination of the wind-induced signals indicate correlations in particular between seismological signals in frequencies above 1 Hz and the local wind obtained at meteorological station MET1. The example of the spectrogram from station DES15 (Figure 6.14), closest to the wind measurements, during Event M1, supports these conclusions. The signals in the spectrogram correlate with the mean wind speed without any visible time lag and they rise in magnitude and bandwidth with increasing wind speed.

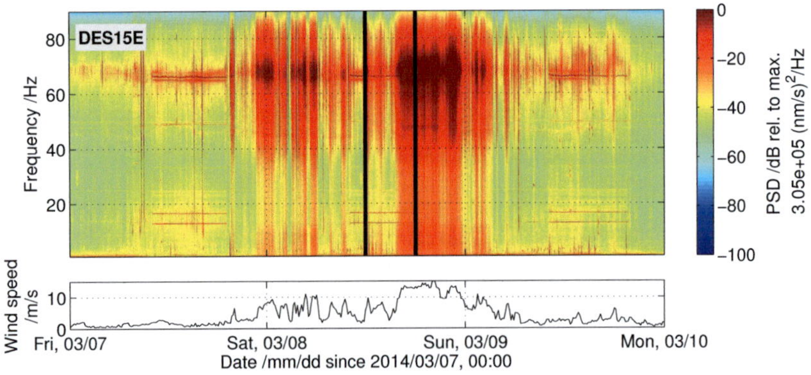

Figure 6.14: Top: Spectrogram of station DES15, E-component from 07.03.2014 to 10.03.2014 for frequencies between 1 Hz and 90 Hz. **Bottom:** Mean wind speed at MET1.

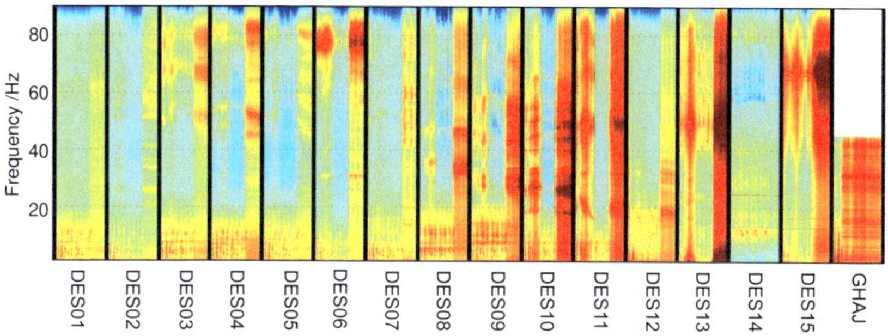

Figure 6.15: Spectrogram samples of all seismological stations between 1 Hz and 90 Hz from 08.03.2014, 12 UTC to 08.03.2014, 18 UTC.

Due to the high local wind speed (cf. Figure 6.10) but no identifiable regional impact (cf. Figure 6.8), Event M1 is well suited for investigating the influence of wind-induced signals from local sources. Therefore, to begin with, the correlation of local wind speed and seismological records is investigated exemplary by M1.

Site Effects

Despite the occurrence of distinct wind-related signals at all stations above 1 Hz, the signal properties at each station, such as the amplitude of the PSD, the frequency content, and the signal shape, differ significantly. The illustration in Figure 6.15 composes samples of spectrograms at all seismological stations for the period from 08.03.2014, 12 UTC to 08.03.2014, 18 UTC (cf. black vertical markers in Figure 6.14), in the presence of a strong local wind speed gradient during M1.

Whereas station DES15 shows the most dominant signal amplitudes in a narrow frequency band around 70 Hz, other stations, such as station DES10, DES11, and DES03 have multiple narrow frequency band streaks of maxima during the presence of strong wind. Other stations, such as GHAJ, show a more uniform stimulation of ground motion, however also interspersed with fine streaks. The ground motion velocity at each recording station seems to be affected very individually despite the close vicinity between most DES stations. The effect of individual ground excitation becomes particularly obvious when comparing the short-period stations DES02 to DES06. They are located at comparable ground conditions in the centre of the array and are presumably exposed to the same wind field, but they still show unique ground excitations of the different frequency bands.

However, the induced signals differ only in terms of frequency content and signal intensity. The temporal occurrence is the same at each station. The distinct, unique streaks moreover indicate that each station site has an individual, specific set of eigenfrequencies, at which the ground at each site has resonances. It is impossible to distinguish, whether the specific characteristics derive from the station underground itself or from close obstacles. At stations DES15 and DES13 buildings, such as a military station, and the Panoramic Complex, respectively, the meteorological tower of MET1 can be considered as potential stimulators. The sites of the other stations are more uniform without any nearby obstacles except some small rocks close by, that might be considered. According to professional experience of a colleague[1] working in the DRRISWIV project[2], similar, unique, and site specific signals associated to wind can be found at seismological stations deployed in the Antarctica, buried below even terrain.

The signals at each seismological station have also uniform properties that are unlikely to be explained by different local wind fields. More likely, site effects, depending on the individual ground surface touched by wind, the individual composition of the shallow ground close to the stations, and the topography cause specific fingerprints in seismological recordings at each station.

[1] Anja Dietz (anja.diez@npolar.no), personal communication at the DGG conference on 16.03.2016 in Münster

[2] Passive-source seismic experiment on the Ross Ice Shelf in West Antarctica, 2014-2016

Correlation of Ground Motion Velocity with Local Wind

The ground motion velocities above 1 Hz are strongly affected by wind, and the wind-induced signals seem to be dominated by individual responses of the sites. However, it is still possible to draw conclusions between the overall wind field close to the stations and the seismological signals. Disregarding the specific frequencies and averaging the PSD from the spectrograms according to Chapter 5.2.3 for the whole frequency band from 1 Hz to the upper frequency limit results in time-dependent PSD magnitude vectors that blur the specific resonance frequencies and make the signals at each station comparable to the local wind field.

To quantify the coherence of two signals cross correlations are a possible choice. The normalised correlation of the local wind speed and the seismological records influenced by the wind field can be assessed by cross correlating the PSD vectors of the ground motion velocity at each station with the local mean wind speed vector. Figure 6.16 shows averaged PSD vectors for all seismological stations, including station GHAJ, for three days beginning 07.03.2014. The mean wind speed at MET1 is given for the same period. At all stations the PSD vectors show high resemblance in their patterns, in particular noticeable for the whole day on 08.03.2014, where peaks appear simultaneously and a period of high values beginning 15 UTC can be identified. All of the graphs, except the PSD at station DES14, show remarkable coincidence with the shape of the mean wind speed vector without any noticeable time shift. The cross correlations of each PSD vector with the mean wind speed (right panel in Figure 6.16) result in high values between 0.5 and 0.8 at DES15, whereas the average cross correlation coefficient of all DES stations is 0.66. The cross correlation coefficient of 0.2 at DES14 thereby is an exception, arising from the high anthropogenic noise level superposing wind-related signals (cf. Figure 6.12). At the more distant station GHAJ, where in all likelihood the mean wind speed at MET1 does not represent the actual local wind field, the PSD

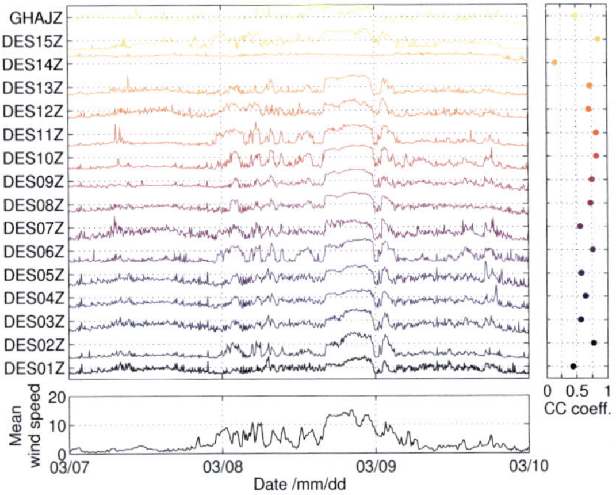

Figure 6.16: Top, left: PSD averaged for all frequencies above 1 Hz with respect to time for all seismological stations for Event M1. **Bottom, left:** Mean wind speed at MET1. **Right:** Normalised cross correlation coefficients of mean wind speed with PSD at each station.

vector still has resemblance with the mean wind speed at MET1 resulting in a cross correlation coefficient of about 0.5.

The power time series in Figure 6.16 at all recording seismological stations itself show high correlations. The auto correlation matrix in Figure 6.17 attests the resemblance. It illustrates the cross correlation coefficients of the power time series at each station with each one another. Power time series of stations DES01 to DES13 have mostly high cross correlations coefficients between 0.55 and 0.91. Station DES15 shows lower correlation values, probably due to its deviant location at higher elevation and the vicinity to obstacles, such as the meteorological tower or buildings of the Panoramic Complex. Station GHAJ has the lowest cross correlations with recordings from the DES array, obviously to ascribe to the distance between array and GHAJ. One would expect a significant trend of correlations depending on the inter station distance, revealing higher correlations for closer stations. However, except for generally higher values between the stations DES01 to DES13 below the emerging cliff at the slope of the valley east of station DES13 (Figure 4.2), there are no higher correlations for closer stations. It implies, that the rugged terrain and the topography have a strong influence, superposing any effects of distance.

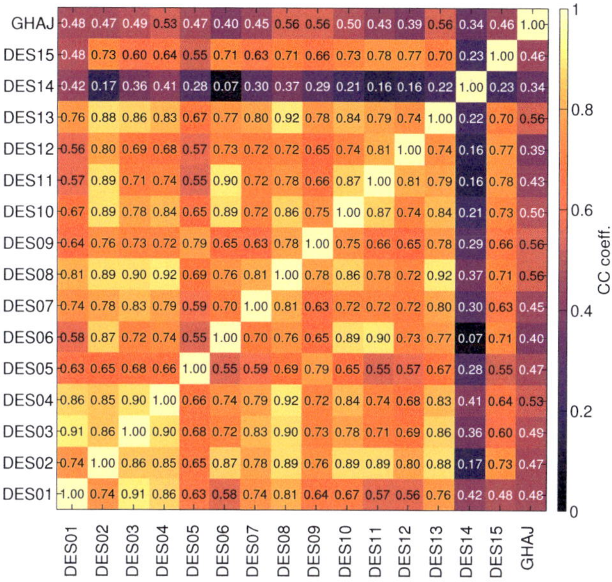

Figure 6.17: Auto correlation matrix of normalised power time series at all seismological stations for event M1. Foundation are power time series displayed in Figure 6.16.

Altogether, the correlation of PSD vectors for frequencies above 1 Hz with the mean wind speed confirm and quantify the visual impression of coherence of local wind speed and ground motion velocity above 1 Hz. The assumption of local wind records representing the influence of wind on ground motion velocity might be limited when it comes to events with higher impact that can not only be described by local winds. As described in Section 6.1.1 Event J1 from 05.01.2015 to 14.01.2015 excels by local wind speed that is increased for a long period. It is a more regional event with high wind speed in a larger area (cf. Figure 6.2) and with an apparent impact on seismological records in frequencies below 1 Hz (Figure 6.7).

As for the previous example, Figures 6.18 and 6.19 illustrate the power time series of the ground motion velocity at all seismological stations. They show strongly matching patterns and therefore despite different sites, they are exposed to similar power spectral densities in the wave field. For Event J1 the local wind speed at MET1 shows accordance with the PSD vectors of the seis-

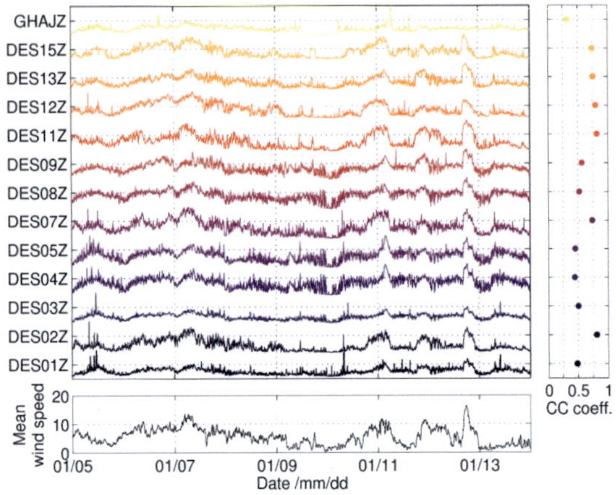

Figure 6.18: Top, left: PSD averaged for all frequencies above 1 Hz with respect to time for all seismological stations for Event J1. **Bottom, left:** Mean wind speed at MET1. **Right:** Cross correlation coefficients of mean wind speed with PSD at each station.

mological records, especially peaks of wind speed, for example on 11.01.2015, 12.01.2015, and 13.01.2015 can be identified in the seismological records, too. The general progressions of the PSD curves is not as much in coincidence with the wind vector as during M1, which is confirmed by the cross correlation coefficients between 0.3 and 0.8 at all stations. The average cross correlation coefficient is 0.60 for the DES stations, about 0.06 lower than during the more local Event M1.

Cross correlation coefficients indicate the degree of correlation of two vectors, ranging from -1 (anticorrelation) via 0 (missing correlation) up to 1 (total positive correlation). Depending on the application, correlation coefficients of more than 0.5 resemble minor correlations, coefficients exceeding 0.75 represent good correlations. For data vectors from totally different origins such as a wind speed vector and the PSD of ground motion velocity from a seismological station, such correlation coefficients of more than 0.5 are remarkable and a clear indicator for a connection. Here, it verifies that the local wind has a significant impact on seismological records in broad frequency bands above 1 Hz and that from the course of the PSD vectors it is even possible to estimate the relative change of local wind speed.

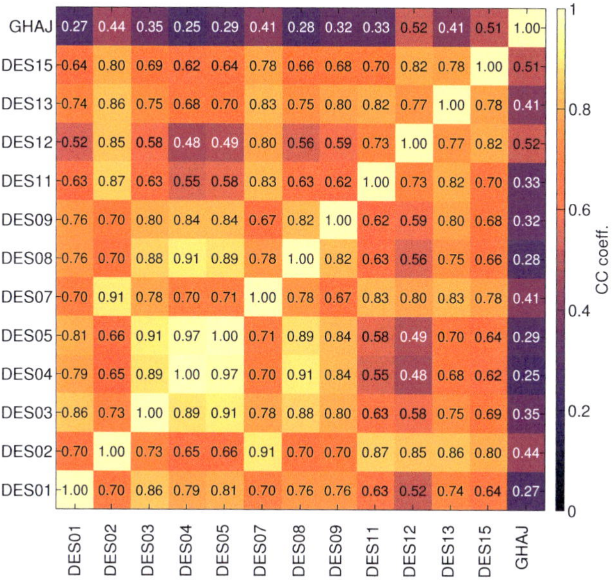

Figure 6.19: Auto correlation matrix of normalised power time series at all seismological stations for event J1. Foundation are power time series displayed in Figure 6.18.

When comparing the correlations during the local Event M1 and the strong Event J1, at first glance, the lower cross correlation coefficient during J1 may resemble a lower influence of wind on the seismological records. However, it should be taken into account that the consideration of in-situ measurements of wind speed only as a factor for the impact of wind on seismological records might be insufficient. Seismological records contain information from far distances as well. It is the local wind recordings that do not describe the wind field in a wider region.

The magnitude of the correlation of the ground motion velocity with the wind speed at MET1 also might depend on the site and the location of the seismological station. The map in Figure 6.20 shows the cross correlation coefficients for each seismological station during Events M1, M2, and J1. The highest cross correlation coefficients well above 0.7 can be found for station DES15 close to MET1. Similar values are reached at station DES12, at the most east-

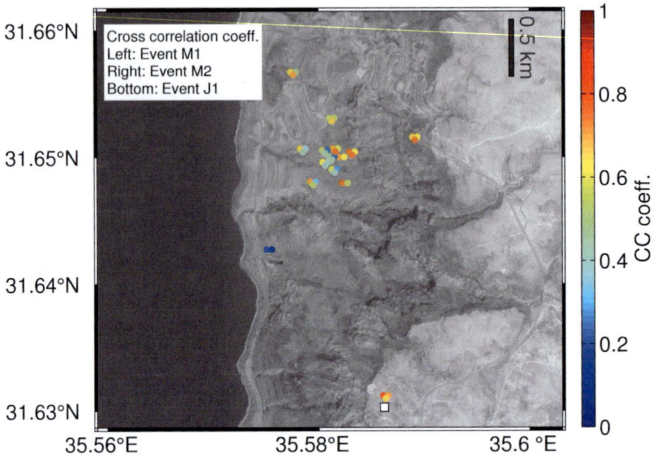

Figure 6.20: Map of the study area. Coloured circle triplets at the locations of the DES stations indicate cross correlation coefficients for the case studies. Top right: event M1, top left:, Event M2, bottom: Event J1. White square: meteorological station MET1.

ern position, also placed at nearly the same elevation as MET1 and therefore presumably exposed to a similar wind field. In the centre of the array, there is a gradient of cross correlation values from west to east, respectively from lower elevations to higher elevations. stations DES02, 06, 08, and 11 are located at higher elevations compared to the remaining stations accompanied with higher correlations of the PSD with the wind speed at MET1 for all events.
The strong depression of the Dead Sea valley involves strong local wind speed differences. At similar elevations one might expect similar wind conditions. Observing the correlation coefficients of ground motion velocity and wind speed with respect to the station locations confirms this assumption. Similar elevations seem to be more important for comparable wind conditions than horizontal distances. stations DES12 and DES15 at similar elevations have the highest cross correlation coefficients, though one station, DES15 is located as near as 100 m from MET1, while the other station DES12 is about 2.5 km distant.

Conclusions

The influence of wind on seismological records above 1 Hz is unique for each seismological station. Wind measurements from a close meteorological station can not be related to the individual excitation of specific frequency bands at each station (cf. Figure 6.15). Local wind affects a broad frequency band of seismological records and seems to excite specific eigenfrequencies that can not be related to the wind properties but to the site characteristics of the stations.

When observing not the specific frequencies at each station, but the whole frequency domain, strong correlations of the PSD of the ground motion velocity and the local wind speed can be found. The magnitude of correlation is highest for high local winds without the superposition of wind influence from farer distances. The differences of the magnitude of correlation at the individual stations can be explained by different levels of station elevation even though one can not assume to expect the same wind field as at MET1.

Seismological recordings in frequencies above 1 Hz are strongly affected by local wind and the wind induced signals in the PSD provide fairly good estimations on the course of local wind field. The magnitude of impact of wind on the seismological records is expected to depend on the presence of obstacles or the influence of topography, such as the flank of the Dead Sea Valley.

6.3.2 Signals from Distant Sources

Spectrograms in Section 6.1.1 and 6.2 with M1, M2, and J1 demonstrate that wind has an effect on seismological records within a broad frequency band and that the effect differs depending on the frequency domain. The example of Event M1, where the local wind field dominates the seismological records (cf. Section 6.3.1), shows that high local wind speed correlates with the PSD of seismic ground velocities especially in frequencies above 1 Hz. Beyond that, the data indicate that the impact of local wind on the ground motion velocities is prevalently limited to frequencies above 1 Hz having only minor effects on frequencies between 0.1 Hz and 1 Hz. However, wind from farer distances, striking big water bodies such as the Mediterranean Sea, has a well identifiable effect on seismological records in lower frequencies.

Figure 6.21 compares PSD estimates for Events J1 (a) and M1 (b), periods with opposing characteristic wind fields and the effect on seismological records, for a broad frequency band between 10^{-3} Hz and 100 Hz. As reference, a period from 01.03.2014 without significant local or regional wind is displayed as well (c). The spectra show PSD estimates averaged for all DES stations, separately for recording components Z, N, and E. The PSD estimates for the GEOFON station GHAJ are given separately. Note, that the strong decrease in amplitude of all graphs below 0.005 Hz is due to the filtering of the seismological waveforms (cf. Section 5.1.1) and needs to be ignored in terms of physical interpretation.

The striking, narrow peak at 0.2 Hz during Event J1, abruptly dropping towards lower frequencies and smoothly decreasing towards higher frequencies, can be directly related to the excitation of a large water body under the influence of high wind speed. In this study, it can be identified as the Mediterranean Sea in a distance of about 100 km, possibly superposed by the influence of the Dead Sea. GFS model analysis for Event J1 confirms the high wind speed over the whole eastern Mediterranean Sea (Figure 6.2). In comparison, during Event M1, when the levels of the PSD graphs in frequencies below 0.1 Hz are similar to J1, and above 1 Hz the levels are higher than during J1, the peak related to ocean microseisms is hard to identify. Obviously, a potential influence of the ocean microseisms is minor compared to the otherwise high-levelled PSD due to local wind.

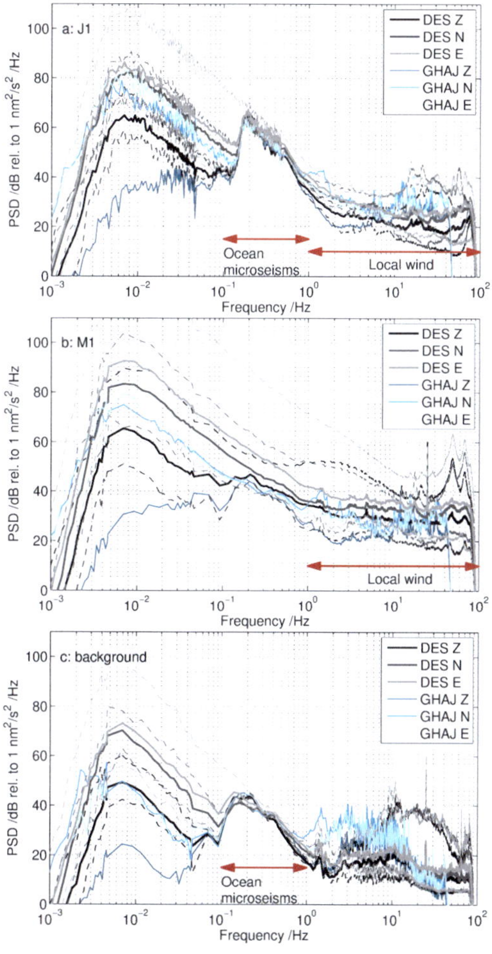

Figure 6.21: PSD at all seismological broad band stations for frequencies between 10^{-3} Hz and 10^{2} Hz. Solid black and grey lines: average PSD of broad band DES stations for Z, N, and E-components. Dashed lines: minimum and maximum PSD values of broad band DES stations. Blue lines: separate PSD estimates of station GHAJ Z, N, and E component. **a:** period during J1 from 07.01.2015, 00 UTC to 07.01.2015, 06 UTC. **b:** period during M1 from 08.03.2014, 18 UTC to 09.03.2014, 00 UTC. **c:** period without significant wind from 01.03.2014, 18 UTC to 02.03.2014, 00 UTC.

In the time period without high wind speed (c: background) in the study area and the surrounding region on 03.01.2014 the overall PSD levels are decreased by 10 dB to 20 dB. The peak at 0.2 Hz, identified for J1, can be observed here, too, although on a level more than 20 dB lower and its maximum less sharply restricted to a narrow peak. This observation indicates that there is always an influence of ocean microseisms, also in calm wind regimes and when the overall noise level is low enough it is detectable.
Observations of seismological records from the DES array and station GHAJ in similar distance to the Mediterranean Sea offer to distinguish whether the signal has its origin only in the Mediterranean Sea or also in the water body of the Dead Sea. The fact, that a strong peak dominates the spectra during a storm with severe wind over the Mediterranean Sea, whereas no peak can be identified during a local storm, when a significant swell at the Dead Sea can be expected too, implies that the influence of the Dead Sea swell plays a minor role despite the vicinity of the stations. Examining spectrograms of ground motion velocity in the frequency domain below 1 Hz along with wind speed records from MET1 and EBS3, in the following section excitations from ocean waves from the Mediterranean Sea and the Dead Sea can be identified and distinguished.

Ocean Microseisms under the Influence of Changing Wind Speed

To identify and understand the strong signals dominating the low frequencies in seismological spectra, it is essential to give attention to the vibration properties of oceans. Ocean microseisms are caused by ocean waves transferring power into the sea floor by releasing compressive stress due to height changes of the centres of mass, fundamentally described by Longuet-Higgins (1950). The main microseismic energy from the oceans is caused by two mechanisms, referred to as primary ocean microseisms and secondary ocean microseisms. Primary microseisms are caused by pressure fluctuations caused by ocean waves hitting the bottom of the ocean. The frequency of the microseisms corresponds to the frequency of the ocean waves. The amplitude exponentially decays with increasing depth of the oceans (Bromirski and Duennebier, 2002), hence shallow oceans cause stronger ocean microseisms.

Secondary microseisms cause a stronger signal, explained by non-linear interaction of two counter-rotating ocean waves with resembling characteristics, such as wavenumber and length. The pressure variations oscillate with the double frequency of ocean waves (Bromirski et al., 2005), not subjected to significant attenuation, and therefore, in seismological records they can be identified as the stronger signal.

According to Kibblewhite and Ewans (1985) secondary ocean microseisms usually can be identified in frequencies between 0.04 Hz and 0.1 Hz in coastal regions. Pierson and Moskowitz (1964) discovered that the oscillation of the ocean waves, respectively, the frequency of the ocean microseisms, afar from the coasts, can be triggered by wind blowing steadily over a large area, eventually reaching an equilibrium with the wind speed. The Moskowitz spectra describe the power spectral density of ocean waves with respect to frequency and wind speed. With increasing wind speed, the oscillation of the waves slows down, and the power spectral density increases coincidentally sharpening the maximum.

For Event J1, for which ocean microseisms are most pronounced, spectrograms of the seismological records in frequencies below 1 Hz can exhibit information on the influence of wind on ocean microseisms in the study area. Figure 6.22 shows an example of two low frequent spectrograms from the broad band station DES01 for the period from 01.01.2015 to 14.01.2015 covering frequencies from 0.07 Hz to 0.1 Hz and 0.1 Hz to 1 Hz. For reasons of comparison, average wind speed records from meteorological stations MET1 and EBS3 are given. When observing signals below 1 Hz, the examination of spectrogram data from one station only is substantially sufficient, the patterns in this frequency domain have uniform shapes for all broad band stations (cf. Figure 6.7).

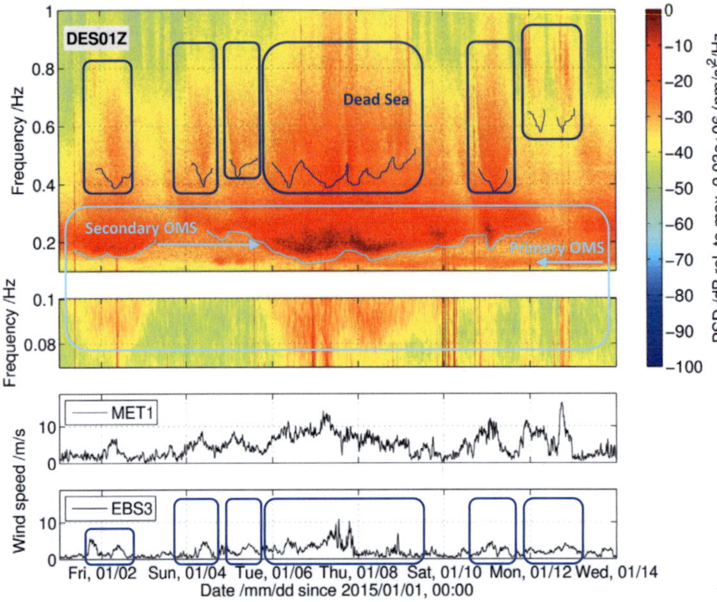

Figure 6.22: Ocean microseisms in spectrograms of station DES01, Z-component, for frequencies between 0.07 Hz and 1 Hz. Average wind speed measurements from MET1 and EBS3 are given with respect to the same time axis. Light blue marks indicate primary and secondary ocean microseisms (OMS). Dark blue marks highlight signals probably representing microseisms from the Dead Sea coinciding with the development of the wind speed at EBS3. The lower frequency limits of the signals, highlighted by coloured lines, are traced by hand.

Two different signal types can be recognised: multiple signals above 0.5 Hz and long, intense signals below. The peak related to ocean microseisms, observed in Figure 6.21, can be identified in the lower part of the upper spectrogram as a sequence of strong signals. The lower level of the sharp maximum of the signal is not constantly located at 0.2 Hz but fluctuates between frequencies from 0.1 Hz to 0.25 Hz, highlighted by the light blue lines in the spectrogram. The main frequency of the ocean microseisms depends on the wind speed measured at MET1. Rising wind speed coincides with a down shift of the main frequency, whereas decreasing wind speed coincides with ocean microseisms shifted towards higher frequencies. Below the strong signal maxima, at about half the frequency, a narrow band and weak second maximum is visible. Like the main maximum, the second signal maximum shifts towards deeper frequencies in the presence of higher wind speed, particularly clear in frequencies below 0.1 Hz, where the lower spectrogram provides a higher frequency resolution.

The signals clearly can be related to weak primary and strong secondary ocean microseisms (see also Figure 6.23), intensified and frequency-shifted downwards by strong winds. However, the comparison of the signals with in-situ measurements from our meteorological stations is insufficient, since the local measurements do not provide any information on the wind field at the signal source, the Mediterranean Sea. In combination with the GFS analysis for the Eastern Mediterranean (cf. Section 6.1.1) the wind speed records from MET1 are nevertheless sufficient for an estimation on the development of the distant wind field.

The other type of signals in the spectrograms in Figure 6.22 is not as unambiguous to classify as the ocean microseisms. The signals above 0.4 Hz, superposed by the upper limits of ocean microseisms, temporally correlate with local wind speed maxima at EBS3, and sometimes MET1. Other than for the ocean microseisms showing only diffused correlations with the local wind speed, the higher frequent signals correlate with wind speed changes on shorter terms. Therefore, it seems likely, that the source is closer to the seismological stations. The fact, that the signals show discrete but conclusive correlations with the wind field at EBS3 at the bottom of the Dead Sea valley, indicates that the water body of the Dead Sea might be the cause.

The perceptions of microseisms from oceans, based on the assumption of large water bodies and not directly conferable to smaller volumes like the Dead Sea, should no longer be considered appropriate. In the Dead Sea valley, the impact of the shorelines of the Dead Sea becomes a more major factor presumably interacting with the influence of wind on different scales. There are only few publications on microseisms from smaller water bodies, however, Lynch (1952) finds microseisms related to the Great Lakes in the U.S. in frequencies of about 0.5 Hz, in higher frequencies than ocean microseisms and well within the frequency domain of the signals in this study. The high density of the Dead Sea salt water of about $1.24\,\mathrm{kg\,l^{-1}}$ (Siebert et al., 2014) might be another factor to consider in terms of wave heights, compressive force, and oscillation frequency.

Compared to the signals of the ocean microseisms from the Mediterranean Sea, the identification of wind induced signals from the Dead Sea is more insecure and the interpretation to handle with caution. However, the frequency domain, the temporal occurrence, and the uniform character at all DES stations, as well as the absence of distinct signals at the more distant station GHAJ (cf. Figure 6.7) are strong indicators for the assumption.

The particle motions from the two types of signals, ocean microseisms from the Mediterranean Sea and from the Dead Sea give information on the dominating wave type stimulated by the signal source. Figure 6.23 shows particle motions in the Z-E plane, corresponding to vertical-horizontal projection in the direction of the propagation for periods of two minutes each. The left panel shows the particle motion during for the presence of secondary ocean micorseisms on 06.01.2015, 23UTC in a frequency band between 0.15 Hz to 0.3 Hz. The particle motion in the right panel comprises a period with a signal related to the Dead Sea microseims on 11.01.2015, 00UTC, filtered for the dominant frequency domain from 0.4 Hz to 0.6 Hz.

The ocean microseisms show vertical and longitudinal wave particle motions, thereby, the vertical motion is stronger pronounced. The wave field is dominated by Rayleigh wave motion, with a strong vertical component. The Rayleigh wave type polarization indicates that the signal is caused by pressure oscillations from interfering ocean waves recorded as secondary microseisms (Matsuzawa et al., 2012). The signal from the Dead Sea waves is also dominated by Rayleigh wave motion, but with a stronger longitudinal component.

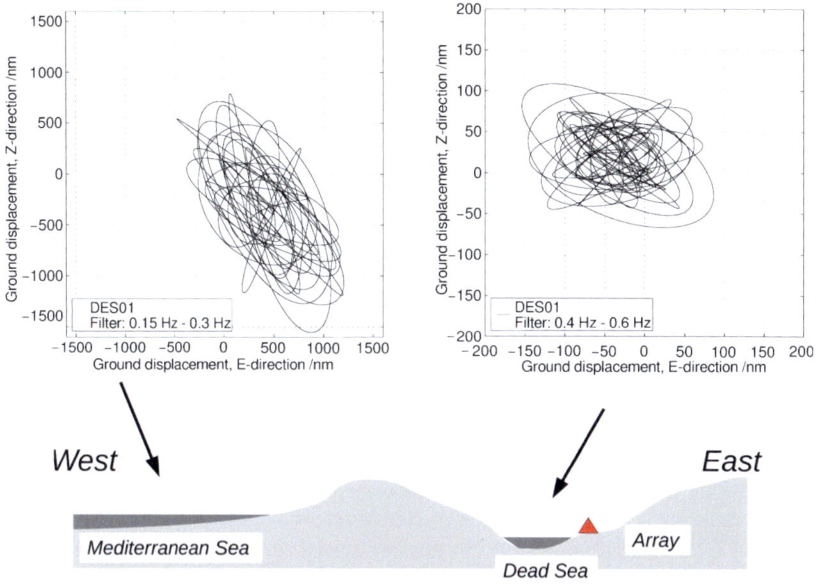

Figure 6.23: Vertical vs. horizontal particle motions in east direction at station DES01 during Event J1 for two frequency domains, where secondary ocean microseisms and potential microseisms from the Dead Sea is identified. **Left:** Period for two minutes on 06.01.2015, 23 UTC, bandpass filtered from 0.15 Hz to 0.3 Hz. **Right:** Period of two minutes on 11.01.2015, 00 UTC, bandpass filtered from 0.4 Hz to 0.6 Hz. **Bottom:** Schematic cross section from west to east across the Dead Sea. The particle motions are assigned to the Mediterranean Sea (top, left) and the Dead Sea (top, right). The sources are both west of the array.

Pressure changes hitting the floor of the lake in 300 m depth below water surface (Hall, 1996) may not be the most dominant part of the source. Instead, coastal waves hitting the eastern and western shores may stimulate the horizontal East-component dominantly.

When interpreting the particle motions, it is essential to consider that only short periods during the signals are observed. In combination with the results from the spectrograms, however, the particle motions substantiate the identification and classification of the microseisms related to the water bodies.

6.4 fk-Analysis on Wind-Induced Signals

Similar as to conduct fk-analysis to identify the origin direction of signals from tectonic ruptures (Chapter 4.1.5), fk-analysis may reveal directed, uniform signals induced by wind and the direction to its source (Groos, 2007). The configuration of the DES array thereby allows to investigate frequencies in the range of:

$$f_{min} = 1.1\ \text{Hz} < f < f_{max} = 19\ \text{Hz}.$$

The restriction of the frequency domain confines the fk-analysis to the frequency domain assigned to local influence of wind (Section 6.3.1). Other than in frequencies below 1 Hz (Section 6.3.2) spectrograms of ground motion velocity in frequencies above 1 Hz reveal non-uniform character at different stations associated with the ambient, turbulent character of wind flow across the ground.

Results of fk-analysis during event M1 confirms the conclusions. Figure 6.24 shows the sliding time window fk-analysis of the DES array for a period of three days from 07.03.2014 to 10.03.2014 in comparison to the wind speed recorded at MET1. The length of the time window of each calculation is chosen as 140 s, corresponding to the duration wind of 20 m s^{-1} needs to cross the complete array. The absolute power of the array signal approximately scales with the wind speed at MET1. Despite the apparent effect of wind on the power of the ground motion velocity, no organisation of backazimuth and slowness values can be identified. The fk-analysis does not reveal any unambiguous identification of predominant signal direction or phase velocity.

The fk-analysis can confirm the ambient character of wind-induced seismic signals, as investigated on the basis of spectrograms of ground motion velocity in Section 6.3.1. In this context, it must be further considered that seismic signals related to the local wind field may not be approximated as a planar wave front crossing the array (Chapter 2.3, Almendros et al., 1999). It is possible, that this assumption is not complied, prohibiting the analysis using fk.

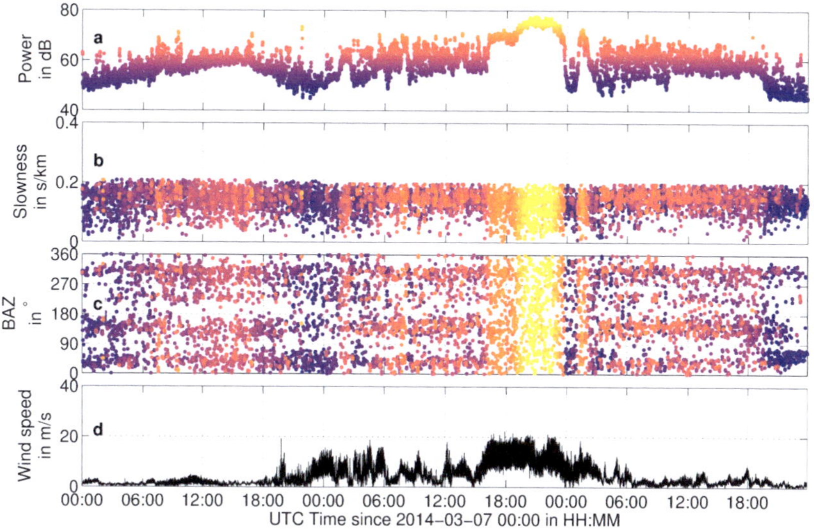

Figure 6.24: Sliding time window fk-analysis of all DES stations except DES14. Length of sliding time window: 140 s, Time steps: 30 s, frequency domain: 2 Hz - 10 Hz. All panels are displayed with respect to the same time axis. Colours of the dots scale with the power in Panel a. **a:** Absolute power. **b:** Slowness. **c:** Backazimuth **d:** Wind speed recorded at MET1.

Other than local wind, enhanced ocean microseisms due to wind (Section 6.3.2) may comply the conditions for fk-analysis: the character of the signals is more uniform at different stations and the signal sources are more distant, hence the assumption of plane wave fronts crossing the array is satisfied better. However, the geometry of the DES array of small aperture is not sensitive to frequencies below 1 Hz. Arrays of larger extent may be capable to track the direction of storm sources and to characterise dominating phases.

6.5 Impact of Wind Related Signals on Data Quality

In seismology, there can be a broad variety of signals in a wide frequency band, from the free oscillation of the earth at frequencies in the domain of mHz (Masters and Widmer, 1995) to local tectonic events reaching frequencies of hundreds of Hz, all subjected to investigations based on seismological measurements. In fact, several research branches even exploit the presence and distribution of ambient noise in seismological records for noise interferometry to gain information on the structure of the ground (Draganov et al., 2007; Shapiro and Campillo, 2004). The influence of wind on seismological records, respectively wind related signals, usually does not belong to the category of desired signals. They affect a broad frequency band, are very site specific, and with varying magnitudes they superpose with desired signals. It not possible to determine universal predictions on the influence of wind, since it depends on many site specific factors, such as the distance to obstacles and the type of obstacles, as well as dynamic factors, such as the meteorological conditions. Chapter 7 determines an empirical approach to relate the influence of wind in terms of local wind speed to ground motion velocity in the study area. This section illustrates and opposes the influence of wind events from the case studies in relation to selected earthquakes.

Previous sections discuss wind related signals on the base of spectrograms of power spectral densities of ground motion velocity for narrow time windows with respect to the frequency. The original data exists as ground motion velocity in the time domain, the base for all further calculations. The wind related signals can already be identified there. Figure 6.25 shows time series of all seismological DES stations during five hours of the local Event M1, when significant changes of the wind speed can be observed at MET1. At all stations the amplitudes of the ground motion velocity increase with increasing wind speed. Below $10\,\mathrm{m\,s^{-1}}$ wind speed only the noise level at station DES14 is high. Figure 6.25 opposes time series of all seismological stations, illustrating that the impact of wind on each of the station differs in amplitude. Note, that due to site characteristics the noise level at all stations is different, but all time series are displayed with respect to the same ordinate without scaling. Therefore, the absolute amplitudes differ, and at stations such as DES01 and DES02 the increase of amplitude with rising wind speed is barely visible.

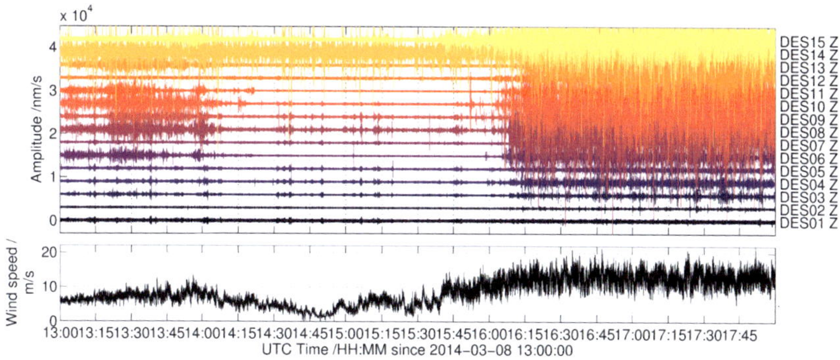

Figure 6.25: Ground motion velocities at DES01 to DES15, Z-components, from 08.03.2014, 13 UTC to 08.03.2014, 18 UTC and wind speed at 10 Hz sampling rate at MET1. Applied filter: High-pass 0.2 Hz

The time window shown here represents a five-hour excerpt, the actual signal, respectively the succession of the signals as a swarm, lasts for more than one day in the case of Event M1. Unlike common earthquakes, the signals related to wind are not caused by abrupt ruptures and therefore do not cause distinct signals with unambiguous phase arrivals and specific durations. In shape and variability, in the time domain, they bear resemblance to volcanic or tectonic tremor signals (see for example Shelly et al., 2007).

Figure 6.26 contrasts the time series from Figure 6.25 under the influence of wind with a recording period of the same duration, containing a small, regional earthquake of Magnitude M=2.9 in 100 km epicentral distance[3] to the stations. The earthquake signal, at 20:12:40 UTC marked by a red frame, has a duration of about 90 s and is barely visible on the long time scale. In more adequate, shorter time windows (cf. Figure 6.27) the earthquake is easier to identify, the time scale here is chosen as to give an impression on the time scale differences of earthquakes and wind related signals. The amplitude of the earthquake signal can be identified, but it is smaller than various other signals within the time

[3] Source: Earthquake catalogue of the Jordan Seismology Observatory

window. More important, the amplitude of the wind related signal is up to ten times stronger than the amplitude of the earthquake at some of the stations.
The characteristics of earthquake signals in seismological records highly depend not only on the magnitude of the earthquake but also on the epicentral distance to the source. To give a more thorough impression on the impact of earthquakes of different type on ground motion velocities in this study, Figure 6.27 shows time series and spectra of three earthquakes, recorded at station DES15 with the Z-component: a small local earthquake[4], the regional earthquake of M=2.9, shown in Figure 6.26, and a teleseismic earthquake[5] of M=6.9 near the Agean Sea in 1330 km epicentral distance to station DES15.

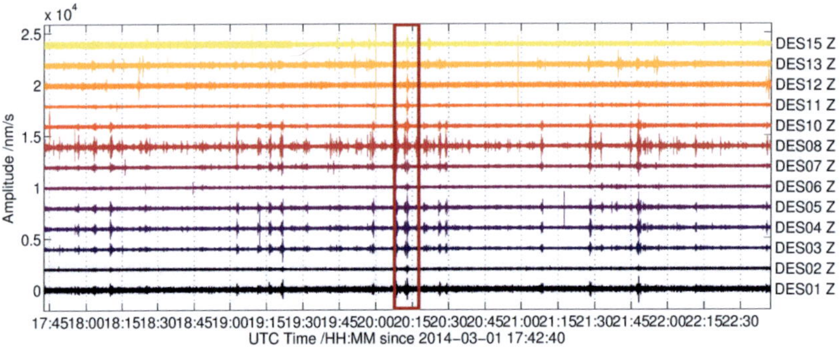

Figure 6.26: Ground motion velocities at DES01 to DES15, Z-components, from 01.03.2014, 18:12 UTC to 01.03.2014, 22:12 UTC. Applied filter: High-pass 0.2 Hz

[4] Located by Gesterman et al. (2015) of Magnitude M=2 with an epicentral distance of about 12 km in the scope of CTBTO test measurements in Jordan

[5] Source: earthquake catalogue of EMSC (European Mediterranean Seismological Centre)

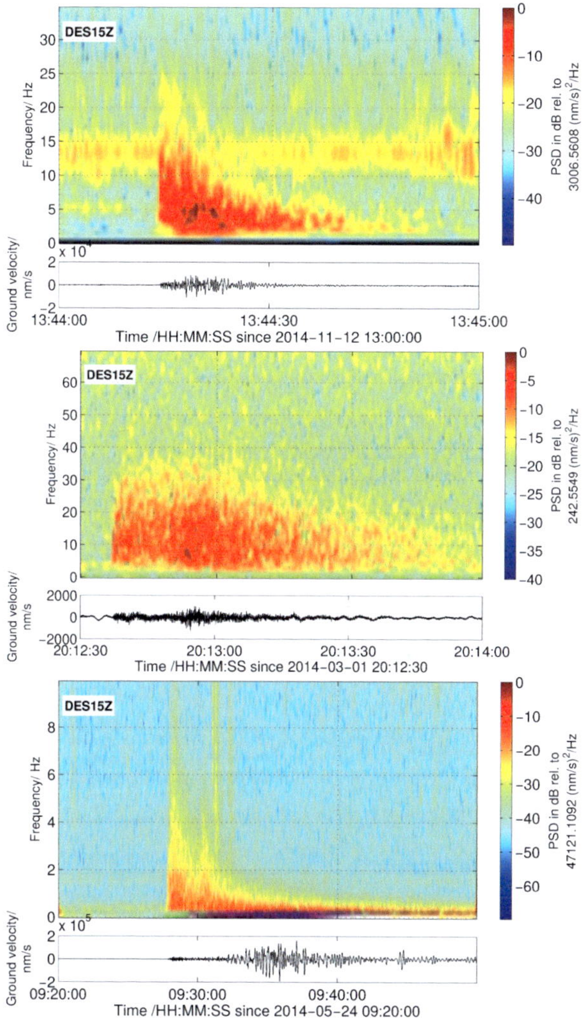

Figure 6.27: Spectrograms of three different earthquakes recorded at station DES15, Z-component for varying time windows and frequency domains. **Top:** Earthquake with magnitude M=2 with an epicentral distance of 12 km from the station. Filter: high-pass 0.02 Hz. **Middle:** Regional earthquake with magnitude M=2.9 in an epicentral distance of 100 km. Filter: high-pass 0.02 Hz. **Bottom:** Earthquake with magnitude M=6.9 and epicentral distance of 1330 km. Filter: high-pass 0.001 Hz.

The spectrograms reveal distinct signals for all earthquakes on different temporal scales and frequency domains. The amplitude and the frequency domains not only depend on the magnitude of the earthquakes, but also on the distance of the sources. The smallest earthquake with a magnitude of M=2 shows a more distinct signal and a higher power spectral density maximum than the M=2.9 earthquake due to its vicinity to the recording station. Compared to the other earthquakes, the teleseismic event has the highest power spectral densities at DES15 during the arrival of the surface waves after 09:30 UTC, but the frequency band of the signal reaches only up to 10 Hz due to the stronger attenuation of the high frequencies diminishing along the long travel path. The large earthquake signal is visible in the time series for more than 40 min, compared to the smallest earthquake with a duration of less than 60 s.

Intensity of Wind Related Signals in Relation to Earthquakes

The duration, the intensity, and the frequency domain of the earthquakes are highly variable, and other signals, or high noise level of seismological records can superpose desired signals. If desired signals and noise do not meet the same frequency domain, data can be filtered to achieve better signal-to-noise ratios, although it is to prefer not to use strong filters, since they may change the phase and amplitude onsets of the seismic phases. When signal and noise meet the same frequency domain, filtering becomes futile.

In order to estimate the impact of wind on seismological records at one station and to assess which type of tectonic signals can be affected by wind induced noise, a noise model in terms of spectra can be de developed. Figure 6.28 comprises the spectra of presented earthquakes (red graphs) and wind signals from Events M1 and J1 (black and dark grey), as well as for conditions without significant wind (light grey) for station DES15, Z-component. The comparison of the spectra depending on the frequency domain reveals highly different characteristics. The problem with the influence of strong wind on seismological records becomes obvious: depending on its charactersistics, it affects the noise

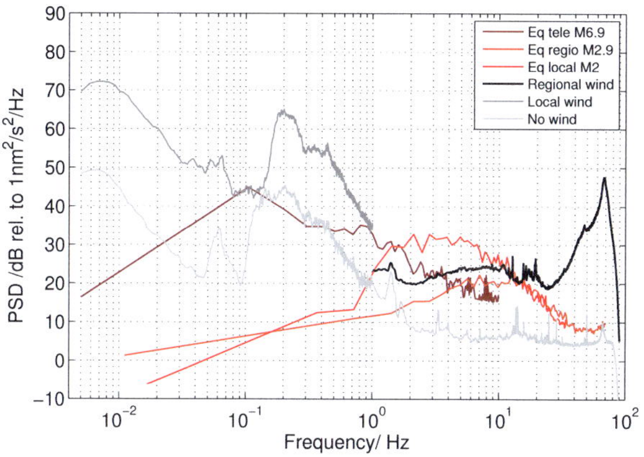

Figure 6.28: Comparison of spectra of earthquakes (Eq) with different magnitudes and epicentre distances with wind related signal at station DES15, Z-component. Regional wind: period from Event J1, between 06.01.2015 and 08.01.2015 with mean wind speed of up to 25 m s^{-1} over the Mediterranean Sea. Local wind: from Event M1, between 08.03.2014 and 09.03.2014, local mean wind speed up to 20 m, s^{-1}. No wind: period from 01.03.2014 to 02.03.2014, mean wind speed below 5 m s^{-1}.

level of the data in the whole frequency domain between 0.005 Hz and 100 Hz, shifting the amplitude level about 20 dB upwards. Signals from local and regional earthquakes, most dominant in frequencies above 1 Hz are superposed by local wind, the black graph of the regional wind partially exceeding the PSD level of the earthquakes. The teleseismic earthquake (dark red) is excelled by the PSD of the local wind in frequencies above 3 Hz, below, the influence of regional wind is up to 20 dB higher.

Figure 6.28 shows an exemplary approach for one seismological station and specific, selected signals, clarifying that the influence of wind on seismological records can have a serious impact on the data quality. It shows, that tectonic signals in the data are not only superposed by wind related signals, but often even weaker. Depending on whether high wind speed is present in vicinity of the recording stations or far away, acting on the seas, the spectrum of wind related noise can vary in shape and frequency domain (Figures 6.21 and 6.22).

In this chapter the influence of wind on ground motion velocity is investigated by several case studies to give estimations and to investigate the sources that are excited by shear stress from the wind. The results are limited to short periods of time and to distinct, high levels of wind speed. A more systematic approach on how different levels of wind speed affect seismological for longer periods is described and analysed in Chapter 7.

7 Scaling of Power Spectral Density of Ground Motion with Wind Speed

Seismological waveforms offer a vast range of research applications, from earthquake seismology to the tomography of the earth's interior, or oceanic microseisms. Depending on the purpose, seismological records need to meet specific signal-to-noise ratios in different frequency domains. Therefore, it is essential to debate the quality of seismological records influenced by various external sources, such as the omnipresent wind field. Chapter 6 demonstrates the effect of wind as signals in seismological recordings. To estimate or predict the significance of seismological recordings in the presence of a dynamic wind regime, a more systematic approach is presented in the following.

The winter season from 01.11.2014 to 31.01.2015 at the Dead Sea provides a sufficiently long data set with a broad variability in wind speed, with the highest wind speed above $10\,\mathrm{m\,s^{-1}}$ from southeast and southwest (Figure 7.1). The winter season serves here as study period, for which continuous spectro-

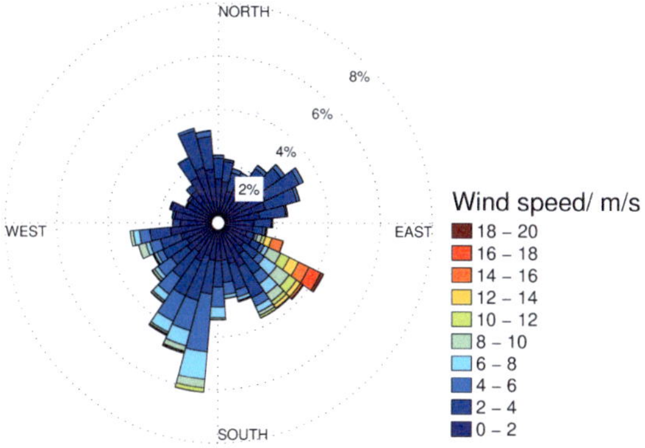

Figure 7.1: Distribution of 10 minute mean wind speed and wind direction at MET1 from 01.11.2014 to 31.01.2015.

grams for all seismological stations are calculated, as exemplary presented in Chapter 6. On the basis of the long datasets, seismological and meteorological records can be collocated and analysed in terms of coherency. On the base of a systematic correlation of the recordings, an empirical relation is defined. Methods, essential to understand the analysis used in this chapter, are specific developments for this study, and therefore explained in Chapter 5.3.

7.1 Susceptibility of PSD of Ground Motion Velocity to Increasing Wind Speed

Other than investigating the impact of wind on seismological records on the basis of single case studies, another approach is to find a more systematic dependency of seismological recordings on wind. The long, continuous meteorological and seismological dataset allows to draw thorough, quantitative conclusions and to develop an empirical relation on the increase of wind speed and the PSD of seismic ground motion velocity. Analysis is conducted for different frequency domains, all seismological recording components, and for all stations separately, as well as altogether as seismic array. The evaluation covers wind from all directions consistently, notwithstanding taking into account any potential influence of the topography, which is subject to Section 7.2.

7.1.1 Change of the Broad Band Noise Level with Wind Speed

Spectra of seismological recordings during different wind regimes described in Chapter 6.5 (cf. Figure 6.27) reveal the strong effect of high wind speed during selected periods on the noise level of seismological records. The analysis of the long continuous dataset from 01.11.2014 to 31.01.2015 enables a more systematic approach, collecting the power spectral density depending on wind speed. Similar to the empirical Moskowitz Spectra, describing the impact of ocean microseisms for different wind speed regimes (Pierson and Moskowitz, 1964), Figure 7.2 illustrates the PSD for 3 continuous months of recordings at all DES Stations, collocated along wind speed steps of $1\,\mathrm{m\,s^{-1}}$.

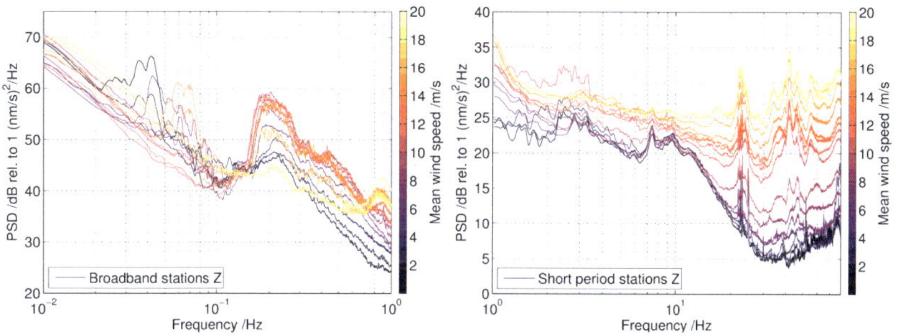

Figure 7.2: Noise level of the DES Array from 01.11.2014 to 31.01.2015 depending on the mean wind speed. **Left:** Average PSD of broad band stations DES01,12,13, and 15 for frequencies between 0.01 Hz and 1 Hz. **Right:** Average PSD of all short period stations DES02 to DES11 for frequencies between 1 Hz and 100 Hz.

The wind speed dependent spectra are calculated according to the description in Chapter 5.3.1 and using time steps depending on the frequency domain as determined by the spectrograms (Table 5.1).

The noise level of the DES array shows a significant, stepwise increase for the whole frequency domain between 0.01 Hz and 100 Hz. The magnitude of the PSD increase highly depends on the frequency band, but within each frequency band the increase of PSD with rising wind speed is predominantly linear. The increase of the PSD with rising wind speed can be detected also for a very low wind speed of 2 m s^{-1}. The comparison of the PSD below 2 m s^{-1} and below 3 m s^{-1} show a slight, but systematic increase for frequencies above 1 Hz, and an even stronger increase of the PSD for frequencies between 0.1 Hz and 1 Hz. The assembly of the noise levels with respect to stepwise wind speed increments shows the magnitude of influence of wind speed on seismological recordings. Other than investigated in the case studies in Chapter 6 and by previous studies (Holub et al., 2008; Ritter and Groos, 2007) not only strong winds above 10 m s^{-1} enhance the noise level of seismological records. The influence of wind can be systematically detected for the whole spectrum of wind speed starting with a level of 2 m s^{-1}, based on an extensive dataset provided by all seismological stations in this study for three recording months. The fact that a comparable impact can not only be found for the temporary DES Stations but also for the permanent Station GHAJ (Figure A.13 in the appendix) implies that the strong effect of the wind on seismological recordings can not be explained by direct exposure of the stations to the wind. The impact of wind in particular also below 5 m s^{-1} on seismological data quality is rather a factor to be reckoned with when estimating the quality of seismological data.

7.1.2 Systematic Dependency of PSD on Wind Speed

In order to quantify the dependency of the PSD of ground motion velocity on the wind speed regime, seismological spectrograms (cf. Figure 6.1.2) can be collocated with mean wind speed data from MET1 and rearranged as histograms. The routine described in Chapter 5.3.2 generates histograms of PSD depending on mean wind speed for all recording components of each seismological station separately.

Figure 7.3 shows results exemplary for the vertical components of the central Station DES01 and Station DES15, closest to the wind measurements. Content of the histograms are data bins extracted from the 3 months spectrograms (cf. Appendix, Figure A.3) in frequencies above 1 Hz, collocated to bins of wind speed, displayed with respect to the abscissa. The ordinate determines the PSD, and the colour indicates the number of data points collocated to each bin. The black graphs left and below the main panel accentuate the distribution of data points per PSD grid, respectively wind speed grid.

Data of both Stations DES01 and DES15 show an overall increase of PSD with increasing wind speed. It confirms the assumption from Figure 7.2 revealing a predominantly linear increase of the PSD with wind speed. The histograms indicate that the PSD ranges within a broad band of approximately 30 dB above a distinct level threshold that increases with the wind speed level (white lines). Both histograms show two main branches of PSD distribution, linearly increasing, and superposing for wind speeds below 5 m s^{-1}, while splitting above. The

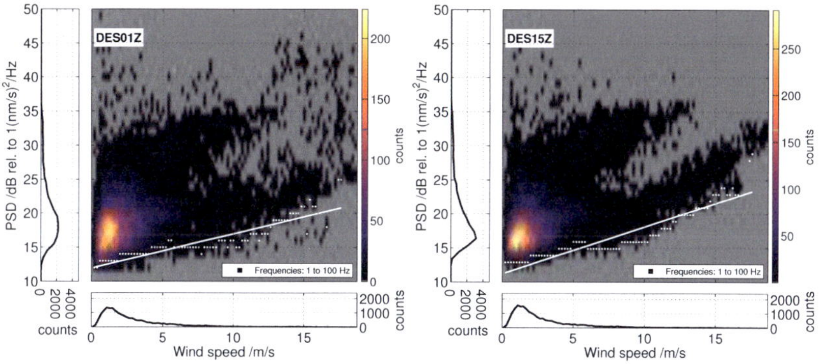

Figure 7.3: PSD in relation to wind speed from 01.11.2014 to 31.01.2015. Frequency domain: 1 Hz to 100 Hz. **Left:** Histogram for Station DES01, Z-component. **Right:** Histogram for Station DES15, Z-component.

lower limit of the lower branch coincides with the lower limit of the histogram. The cause and effect of the splitting is subject to Section 7.2, here the absolute lower limits of the histograms are relevant.

The cloud of data points in the histograms represent the PSD level for three months of recording including the noise level in addition to all kinds of signals such as tectonic events, or anthropogenic noise. All signals contribute to the cloud, however short signals, such as local tectonic earthquakes, have a minor impact, at most contributing few data bins above the lower limit of the histogram. During a comparably long period of three months, tectonic signals would take place without preference in all wind speed regimes consistently. Other than the general level of PSD, that might be slightly increased by the sum of distinct signals contributing to the dataset, the slope of the data points with wind speed is a systematic effect that can be attributed solely to the wind speed.

The lower border of the PSD is of physical relevance, since it marks the PSD level below which, due to the existing noise level, no seismological data at all can be classified. Any signal, from tectonic or arbitrary sources, detected with a certain PSD, i.e. 15 dB, that lies above the lower PSD limit during a calm wind regime should be well within the detection sensibility of the recording station. The same signal, however, would be located below the lower limits during a situation with wind speed above $10\,\mathrm{m\,s^{-1}}$, and therefore superposed by a noise level at the station, which is higher than the signal amplitude itself. Other than the PSD levels in spectra, such as in Figure 6.28, the lower limits do not mark the average noise level of data, but the overall lowest power spectral densities that can be found in the records. Therefore, it is an important indicator for the seismological data quality in general and for the estimation of the detectability of specific signals in particular.

The lower limit of the histogram data is estimated by the 5% percentiles of the bins within each wind speed step of $0.2\,\mathrm{m\,s^{-1}}$ (white marks in Figure 7.3). Compared to the absolute minimum, this value is robust against outliers. The course of the lower limit according to the 5% percentile suggests a weak exponential increase of PSD with wind speed. However, the density of data decays with increasing wind speed, especially above $10\,\mathrm{m\,s^{-1}}$ (cf. black curve in bottom panels of Figure 7.3) and therefore, the increasing slope with higher wind speed has to be interpreted with caution. Further effects, such as the wind direction (cf. Section 7.2) can influence the course of the diagram. Therefore, the cautious, most simple estimation, a linear regression, can describe the dominating trend of PSD increase with wind speed up to $18\,\mathrm{m\,s^{-1}}$ fairly sufficiently (white line in Figure 7.3). Note, that the PSD is represented in units of dB in this study. Physically, a linear increase of the PSD in dB represents an exponential increase of power spectral density along with a linear increase of wind speed.

Figure 7.4 summarizes the lower PSD limits of Z-component ground motion velocity from all seismological stations used in this study for the winter period and in frequencies above 1 Hz. Coloured dots mark the 5% percentiles of the PSD with respect to the wind speed, solid lines in corresponding colours represent the linear regressions to fit the data points. Data from all seismological stations have upward slopes ranging from about $0.2\,\mathrm{dB\,m^{-1}\,s}$ to $1\,\mathrm{dB\,m^{-1}\,s}$, none of the slopes is negative or indifferent.

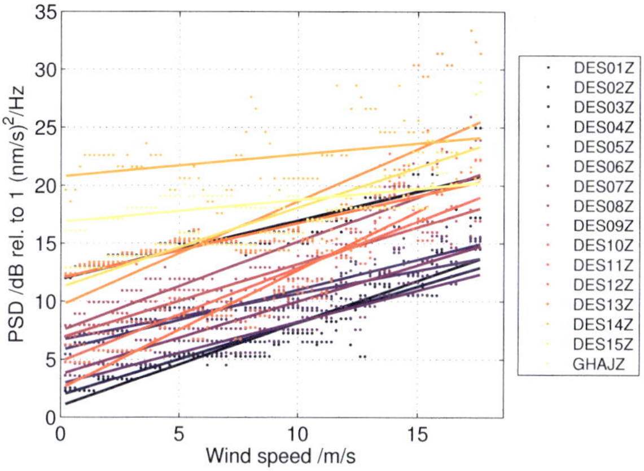

Figure 7.4: Slopes of PSD lower limits with respect to wind speed for frequencies between 1 Hz and 100 Hz for all seismological stations, Z-component. Dots: 5% percentiles of histogram data; solid lines: corresponding linear regressions. Period: 01.11.2014 to 31.01.2015.

Data of all stations show a clear upward trend of power spectral densities with wind speed, and all slopes are well within the same scale. Stations with slightly divergent behaviour are Stations DES14 and GHAJ. DES14 has a slope of only 0.2 dB m^{-1} s, while at the same time the power spectral density has the overall highest level compared to the other stations. The scattered distribution of data points is insufficient for the linear regression and probably caused by the low data quality of the station as discussed in previous chapters. Station GHAJ as a permanent station may be less affected by wind, since it is better isolated against the direct impact of wind by the deeper deployment below surface (cf. Chapter 4.3). A more plausible cause for the less steep slope are the comparably high noise level and the distance of 50 km to the wind measurements. It is likely, that the local wind field dominating the impact on the seismological recordings is different at GHAJ degrading the quality of comparison between wind and ground movements.

The comparison of ground motion velocity with local wind speed measurements for the frequency domain above 1 Hz presumably provides the most sustained results, since this frequency domain best represents the local wind impact. However, the same calculations for the lower frequency domains from 0.01 Hz to 0.1 Hz and 0.1 Hz to 1 Hz result in unambiguous relations between the rise of wind speed and increasing PSD levels. Calculations are conducted for all seismological stations for various different instances, listed in Figure 7.5: three non-overlapping frequency bands, each recording component Z, N, and E, and for two datasets *all* and *nt*. Hereby, *all* represents the whole dataset, *nt* is a reduced dataset covering only the night times from 03 UTC to 15 UTC chosen according to the day times with lowest anthropogenic noise (cf. diurnal variation of noise level Figure 4.10. The dataset *nt* is smaller than the complete dataset but adjusted to exclude the strongest anthropogenic noise. The table in Figure 7.5 lists the slopes of the 5% percentiles, emphasized by a colour bar, for all cases. In addition to the results for the single stations the mean slopes and the standard deviations are listed for all cases at the bottom. Mean and standard deviation are calculated for Stations DES01 to DES14 comprising the comparable stations.

Observing the results decomposed for all cases and stations enables several conclusions. For all single cases, the slopes of the 5% are positive, indicating that an increase of wind speed results in an increasing lower limit of the PSD. It implies that at all seismological stations increased wind speed causes a higher detection threshold of signals in seismological records.

The standard deviations of the mean values are smaller than the mean values, at maximum at about 50% of the mean values. Concerning the diverse station sites at different elevations the errors can be considered small. The slopes in Figure 7.5 at each single station are enhanced for the horizontal components N and E in comparison to the vertical component Z in all frequency domains. Wind induces higher energy in the horizontal recording components than in the vertical recording component supporting the theoretical considerations on frictional stress at the surface (cf. Chapter 2.2).

The reduced datasets omitting day time periods with higher anthropogenic noise provide stronger slopes and slightly enhanced errors. The bigger errors can be explained by a smaller dataset reduced to half size, the stronger slopes are probably the result of the dataset's higher quality. Anthropogenic noise

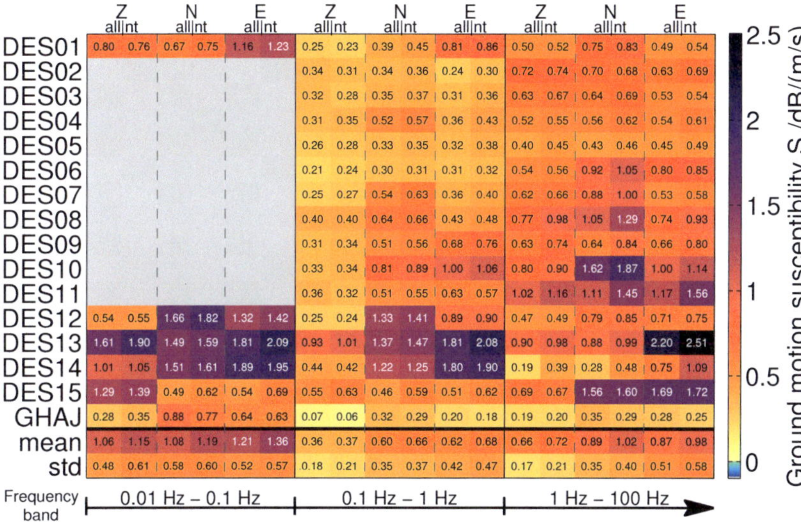

Figure 7.5: Ground motion susceptibility $[S]$=1 dB m^{-1} s for all stations for various frequency ranges and noise conditions. **Top to bottom:** seismological stations, mean, and standard deviation (std). **Left to right:** Frequency bands between 0.01 Hz and 100 Hz (respectively 50 Hz for Station GHAJ). For each frequency band and each station: Z, N, and E component for the complete dataset (*all*) and the reduced dataset comprising only night times (*nt*). Colours match the slope values and serve as clarification. No calculations are conducted for the lowest frequency band for short period stations DES02 to DES11.

superposes signals in similarly broad frequency bands as the influence of wind and it affects the seismological recordings for several hours each day, amounting a significant part of the dataset. Therefore, the slopes of the data during night time provide results for stations with a better data quality. In turn, it means that seismological records with minor data quality due to strong anthropogenic noise is less affected by the influence of wind, however contemplated on a strict relative scale in comparison to the already noisier data.

Figure 7.6 subsumes the results of Figure 7.5 by opposing mean values and standard deviations of all stations for all parameters. It becomes obvious that the horizontal recording components are affected by wind significantly stronger

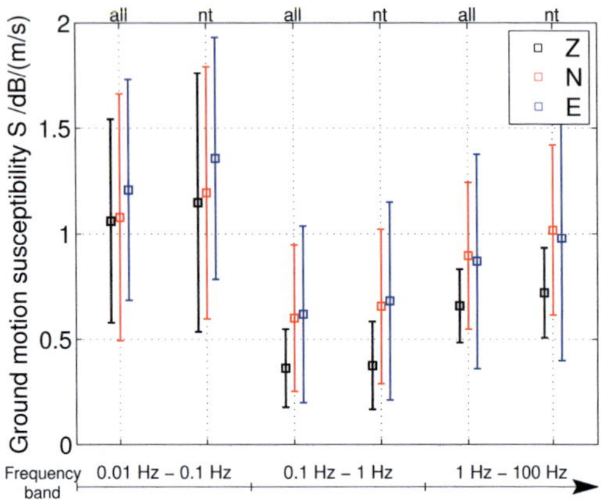

Figure 7.6: Mean ground motion susceptibility for three frequency domains and two datasets *all* and *nt*. Colours indicate recording components, error bars mark the standard deviations.

than the vertical component Z. As expected from the results on ocean microseisms in Chapter 6.3.2 the increase of PSD along with increasing local wind speed is higher in frequencies above 1 Hz than between 0.1 Hz and 1 Hz. In the medium frequency band the influence of ocean microseisms is dominant, and though it is directly linked to strong winds, the observation of wind in-situ only is insufficient to detect correlations.

The high slopes in the lowest frequency band between 0.01 Hz and 0.1 Hz indicate a strong excitement of seismic ground motion velocity in relation with local wind. The magnitude of the slope in the lowest frequency band exceeds the slopes in the other frequency bands by a factor of up to two. As discussed in Chapter 6.3.2 the influence of ocean microseisms fades below 0.1 Hz, as well as the influence of anthropogenic noise (McNamara and Buland, 2004). However, wind seems to have a striking impact on the lowest frequencies. Excitation of ground motion velocity in frequencies between 0.01 Hz and 0.1 Hz can not straightforwardly be explainable by turbulent fluctuations of wind. A direct link between the ground motion velocity and barometric pressure changes is unlikely as well. During the investigation period, the local wind speed and the run of the barometric pressure do not show closely resembling characteristics (cf. barometric pressure and wind speed in Appendix, Figure A.4).

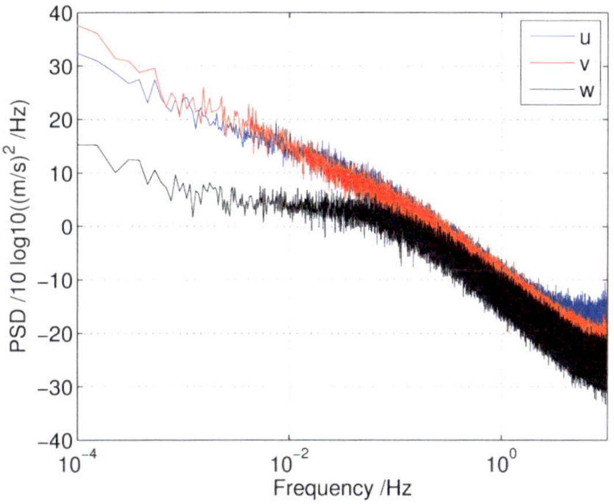

Figure 7.7: PSD of 3-dimensional wind speed from 07.01.2015, 00 UTC to 08.01.2015, 00 UTC at MET1 during event J1.

Still, the strong dependency of power spectral density of low frequent ground motion velocity on wind speed changes is an explicit indication for wind. The strong excitation of seismic ground motion velocity in frequencies below 0.1 Hz thereby thoroughly fits the power spectral density distribution of wind over the frequency. Figure 7.7 reveals the power spectral density of the turbulent wind speed components recorded during event J1 at MET1. The PSD of the wind rises towards lower frequencies attesting a high release of energy into the ground for frequencies between 0.01 Hz and 0.001 Hz. The context illustrated in Figure 7.7 confirm the empirical wind spectra by Van der Hoven (1957).

So far, wind directions are not considered. The influence of wind directions can not be approached as a general issue, it is a strongly site specific parameter, as described in Section 7.2. The study is based on a recordings comprising meteorological events of various scales, durations, and characteristics, including wind from various wind directions. Therefore, despite or precisely because of not considering the wind direction in present results, the systematic, evident relation between the PSD of ground motion velocity and the wind speed is likely to represent a universal correlation widely adoptable. When describing the relation between the logarithmic PSD with rising wind speed the linear regression to the lower PSD level highly satisfies the run of the data and therefore obeys the following principle:

$$PSD_{rel} = S \cdot u_{hor} + e_0, \tag{7.1}$$

where $[PSD_{rel}] = \text{dB}$ represents the power spectral density on the relative, logarithmic scale, as used in all figures in this study, and $[u_{hor}] = \text{m s}^{-1}$ is the horizontal wind speed. $[S] = \text{dB m}^{-1}\,\text{s}$ and $[e_0] = \text{dB}$ are the polynomials. The y-intercept e_0 is individual for each station, recording component, and frequency domain under investigation. It is an indicator for the characteristic data quality of a recording and equals the PSD level during the total absence of wind. Since it is a specific parameter, independent of the wind speed, in this study it is not subject to further interpretation.

The slope S provides a direct interpretation of the effect of wind speed increase on the ground motion velocity. In this study S can be quantified for the DES array between:

$$0.36\,\text{dB m}^{-1}\,\text{s} < S < 1.36\,\text{dB m}^{-1}\,\text{s}. \tag{7.2}$$

Defining the terminology *ground motion susceptibility* S according to Chapter 5.3.3 for $[S] = \text{dB m}^{-1}\,\text{s}$, that describes the increase of the PSD of seismic ground motion velocity per mean horizontal wind speed $[u_{hor}] = \text{m s}^{-1}$ allows to describe and quantify the influence of wind on seismological recordings. The parameter S describes the vulnerability of the seismological recordings to wind for a wide range of wind speeds and depends on station and local ground characteristics. In this study, $S \approx 1\,\text{dB m}^{-1}\,\text{s}$ representing an increase of the noise level of ground motion velocity of 1 dB per $1\,\text{m s}^{-1}$ rise of wind speed. When interpreting the relation between PSD of ground motion velocity and wind speed in a physically more descriptive context, the power spectral density may be converted into the absolute scale and the relation is translated into:

$$PSD = \exp_{10} \frac{S \cdot u_{hor} + e_0}{10}. \tag{7.3}$$

The $[PSD] = \mathrm{nm}^2\,\mathrm{s}^{-2}\,\mathrm{Hz}^{-1}$ describes the power spectral density on an absolute scale and it becomes obvious that the increase of power spectral density actually scales exponentially with rising wind speed. The context described in Equation 7.3 is needed to understand the sensitivity of seismological records to wind speed. However, due to easier handling and seismological conventions, this study will furthermore consider the PSD with respect to the relative, logarithmic scale in dB and stick to the linear relation of the relative PSD with wind speed as described in Equation 7.1.

Yet, there are no publications similar to this study concerned with systematic, quantitative relationships between the wind speed level and seismological records. A case study, where Dietze et al. (2015) compare data from a meteorological and a seismological station, results in a similar linear relation between power spectral density and wind speed confirming the applicability of the results in this study.
The magnitude of the slope S found in this study for the wind excitation of power spectral densities at the DES Station may not have universal validity. However, it gives an estimate of the dimension in which seismological records are affected by wind. More important, the correlation described in Equation 7.1 is stable for the present dataset and specifies a principle describing the relative impact of wind on ground motion velocity. It is therefore expected to have general validity.

7.2 Influence of the Topography on Stress

In the previous section the influence of wind is investigated in quantitative terms of wind speed disregarding the wind direction. Across plain terrain without any obstacles, observing only the wind speed while omitting the wind direction, may be a fairly adequate approach. As described in Chapter 2.2 the dynamic forces acting at the interface between atmosphere and the earth's surface are composed of shear and normal parts depending on the shape of the surface. Over plain terrain horizontal wind causes frictional force described by shear components of the stress tensor. The total amount of shear force thereby depends on the surface characteristics quantified by the coefficient of friction. In the case of obstacles, where wind hits the solid surface with a normal component, the description of the force solely by shear stress is insufficient and needs an additional term for pressure forces (Equation 2.8).
The locations of the seismological stations are chosen as to provide high seismic quality and therefore, with the exceptions of Stations DES13 and DES15 (cf. Table 4.2) they were placed away from local obstacles. The most prominent factors that affect the properties of the wind from different directions are the strong topography in the Dead Sea region and the water bodies of Mediterranean and Dead Sea. The influence of the water bodies is already identified and investigated in Chapter 6.3.2 and can as well clearly be quantified in the specific frequency band below 1 Hz in the seismological recordings (Figures 7.5 and 7.6). To investigate the influence of topography, respectively the wind direction, on seismological records from the DES array, the frequency domain investigated in this section is limited to frequencies above 1 Hz, proven to represent the local wind conditions adequately (cf. Chapter 6.3.1).
When considering topography, the assumption of air flow across a plain surface, where only shear stress acts caused by friction (cf. Chapter 2.2.3), lapses. Instead, strong topography may act similar to an obstacle experiencing upslope flow resulting in additional dynamic pressure forces released at the surface. Figure 7.8 illustrates a very basic idea of air flow along the ground originating

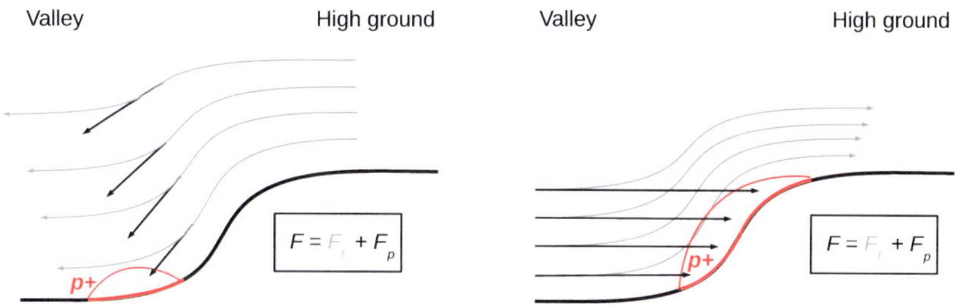

Figure 7.8: Wind-driven forces on the ground for raised terrain. Grey arrows sketch trajectories. Black arrows indicate potential transversal components of the air flow. Other than for plain terrain the acting force may contain dynamic pressure force F_p in addition to friction force F_f. Red areas indicate the surface met by additional, dynamic pressure forces. **Left:** downslope flow. **Right:** upslope flow.

from a high ground into a valley (left panel) and from a valley towards high grounds (right panel).

In both cases find flow along the slopes exists with a deflection from the purely lateral movement (cf. Figure 2.5), bearing also transversal components towards the surface (grey arrows). The equilibrium of contact forces gains a term for the pressure force F_p in addition to the friction force F_f. Transferred to the present scientific question, the stress released by horizontal wind on the earth's surface may increase due to additional surface normal components.

Reducing the problem solely to stress components, however, ignores that the total amount of friction force can be enhanced due to wind speed change. Wind, passing along valleys, or flowing down or upslope, can be subject to the Venturi effect (Cécé et al., 2014) or strong pressure gradients (Koletsis et al., 2009) decelerating or accelerating the wind locally. In the case of Figure 7.8, the downslope wind (left) decelerates, whereas the upslope wind (right) increases upwards (indicated by diverging and converging black arrows), resulting in lower, respectively higher friction forces $F_{f,k}$ and $F_{f,a}$.

The question is, to which amount the tangential flow of air, influenced by the topography, transmits surface stress to the solid ground. Is the input of force into the ground higher for upslope winds than for downslope winds, or approximately comparable? And is it possible to find a systematic influence of topography on the ground motion susceptibility in seismological data?

7.2.1 Friction Velocity as Indicator for Stress at the Surface

In the case of the Dead Sea valley, air flows downslope towards the valley and upslope towards the mountains are of special interest. The question is, how the topography influences the release of flow energy into the ground, and which contact forces dominate. The model in Figure 7.8 may give a good idea, however, probably it describes a vast simplification of processes such as the anyhow exemplary description of wind circulation in the Dead Sea valley in Figure 3.2. In the scope of this thesis, the specific focus is on the western slope of the Dead Sea valley, where the meteorological station MET1 and the seismological stations are deployed.

Meteorological methods yet offer an approach to estimate the wind shear and allow to draw conclusions on the shear stress acting between atmosphere and ground (Chapter 2.2). The turbulent wind components measured at MET1 in 10 m height can be used to derive the friction velocity u_* (Equation 2.6) as a parameter for the shear of the wind field 10 m above ground. Figure 7.9 displays the friction velocity (left panel) and the shear stress (right panel), calculated according to Equations 2.5 to 2.6 with respect to the mean wind speed in 10 m above the ground. The friction velocity increases approximately linear with increasing wind speed. Further, as indicated by the colours, the increase of the friction velocity systematically depends on the wind direction, as emphasized by the linear regressions for the four main wind directions. West wind is subject to the strongest shear, whereas wind from east is influenced least by shear. Wind from north and south are subjected to shear slightly stronger than wind from east, however, there is a poor data base for wind speeds above 5 $m\,s^{-1}$ for those directions (cf. wind rose in Figure 7.1).

The dependency of shear on the wind direction in this case does not imply that the ground surface has different coefficients of friction in varying directions. It

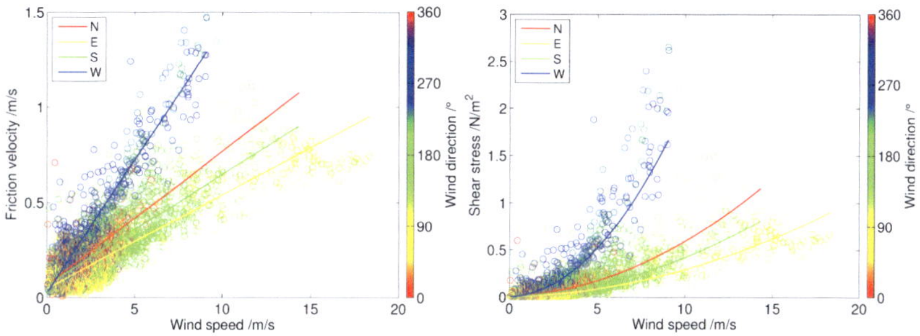

Figure 7.9: Shear parameters depending on horizontal wind speed and direction from 01.11.2014 to 31.01.2015 at MET1. Colours of the circles indicate the wind direction. Solid lines indicate regression for four main wind directions from N ($315° \leq wd < 45°$), E ($45° \leq$ wd $< 135°$), S ($135° \leq$ wd $< 225°$), W ($225° \leq$ wd $< 315°$). The sampling rate of the mean measurements is 30 minutes. **Left:** Friction velocity, regression linear. **Right:** Shear stress, regression squared.

rather shows that the approach to ascribe the shear of the wind in the surface layer to the frictional parameters of the ground is insufficient here. The study area, including the location of MET1, is not an ideal, plain site (Figure 4.15). Instead, the shear of the wind is driven by a rugged terrain, where the strike of the flow towards the steep, upward slope of the Dead Sea valley additionally decelerates the wind and intensifies the wind shear. The effect becomes more apparent when considering the shear stress (cf. Equation 2.5) of the horizontal wind field in Figure 7.9 (right panel). It scales with the square of the friction velocity and therefore, it is even more vulnerable to increasing mean horizontal wind speed. The exceptionally strong increase of shear with wind speed from west can not be explained with the assumption of a stable stratification across the ground any more. It is more likewise to assume additional turbulence of the wind, triggered by additional dynamic pressure applied at the inclined slope. The common model for the calculations of the friction velocity over plain ground does not encompass complex turbulence over elevated terrain. Stull (1988) describes several schematic patterns of wind flow across hills and valleys. Depending on the stratification of the boundary layer, hydraulic jumps may apply, wind circulations can develop, and stationary mountain waves can occur while passing complex terrain (Whiteman, 1990). Air flow from west that passed the Dead Sea valley obviously is subject strong changes in elevation on short distances. Analyses on wind systems inside the Dead Sea valley (c.f. Figure 3.2 Herzog, 2015; Shafir et al., 2008) confirm complex wind circulation and decoupling processes in the boundary layer. Further, surfaces, with different roughness lengths are passed, such as the Dead Sea surface and rugged terrain. The sum of processes expected for wind from west explains the strong wind shear for west wind. More detailed investigations on the uplift of wind by topographical forcing, based for example on the vertical kinematic temperature flux $w'T'$ can not be provided by the meteorological station MET1. However, the analysis of horizontal and vertical variances of the turbulent wind field σ_u^2 and σ_w^2 show both enhanced horizontal and vertical turbulence for wind from west (Appendix, Figure A.15).

Downslope winds, in contrast, produce less wind shear than upslope winds and even the winds along the valley. The downslope winds from east obtained at MET1 pass less complex terrain in both proximate and larger scales (Figure 4.15). They hold less wind shear, possibly due to more stable stratification that may be additionally reinforced by the down flow of the wind (Durran, 1990).

In terms of contact forces at the surface the results for east wind lead to the conclusion that dynamic pressure plays at most a minor role for downslope winds. In addition, friction forces can be assumed weaker compared to winds across plain surfaces, upslope winds, and valley winds. Then, according to model assumptions, no dynamic pressure forces assigned to horizontal wind flow, apply (Figure 7.8).

The interference of the wind field and topography has an impact on seismological records. Similar to the distribution of friction velocity and shear stress of wind with the mean wind speed, the power spectral density of the ground motion velocity scales with the wind speed (cf. Section 7.1). Assigning the PSD of ground velocity with wind speed and wind direction at MET1 results

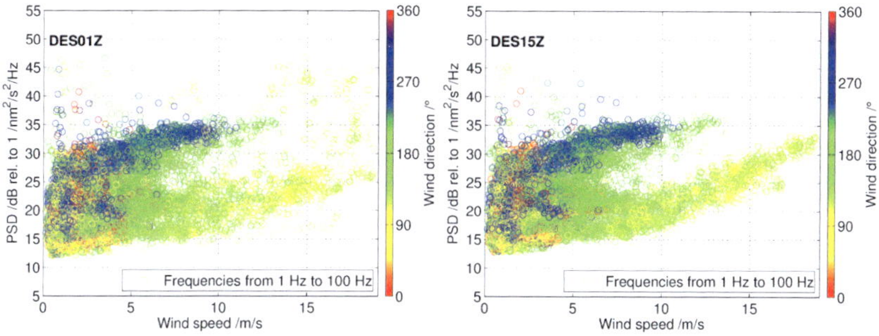

Figure 7.10: PSD of ground motion velocity from 01.11.2014 to 31.01.2015 with respect to mean wind speed at MET1 and coloured according to wind direction. Frequency domain: 1 Hz to 100 Hz. **Left:** Station DES01, Z-component. **Right:** Station DES15, Z-component.

in distributions as shown for Stations DES01 and DES15, Z-component, exemplary in Figure 7.10. The distribution of power spectral density in the scatter plots corresponds to the shapes of the histograms analysed in Figure 7.3, but other than for the gridded data points there, here it is possible to deduce the dependency of the data on the wind direction as well.

In addition to the increase of the power spectral density with wind speed at both stations DES01 and DES15 the two branches of data clouds diverge from a common centroid between 12 dB and 25 dB at low wind speed towards two separate levels at high wind speed. The higher branch is dominated by winds from southwest and west, the lower branch can be assigned to winds solely from east and southeast. Due to the sparsity of high wind speed incidents from other wind directions the according data points disappear in this type of visualization and interpretations of them are not possible. Subsequent illustrations in polar coordinates (Figure 7.11) resolve data for all wind directions steadily. Recordings of stations DES01 and DES15 in Figure 7.10 show remarkable accordance in their patterns. They are selected as examples due to their different locations, one in the centre of the array, and the other one closest to the meteorological station, the highest station of the array. Data of both station show nearly identical distribution of PSD with respect to wind speed and wind direction. Merely the lower slope at DES15 is slightly increased for wind speed above 15 m s compared to DES01. The uniformity of the data argues for the high quality of the seismological records and the site locations. Further, an influence of potential vibrations from the meteorological tower close to DES15 can be classified as imperceptible for the objectives of this study.

When comparing the shear of the wind field in Figure 7.9 with the distribution of the PSD in Figure 7.10 the remarkable qualitative accordance attracts attention. The distribution of friction velocity with wind speed and wind direction largely confirms the gradient of the PSD with respect to the same parameters. It is to anticipate, that with increasing wind speed, friction velocity and shear stress of the wind field increase. Just as well, from hitherto case studies it is expected that energy of ground motion increases with wind speed. Striking here is, that friction velocity, respectively shear stress, and the PSD of ground motion velocity respond not only to increasing wind speed in comparable manners, their resemblance is as high as to react on the wind directions the same way. Both datasets allow to draw the same conclusions on the influence of topography on shear, respectively the energy that is released at the ground surface. The increases of both the parameters of the friction velocity and the PSD of ground motion velocity with wind speed can be compared qualitatively. However, quantitatively, they take place at different scales and for unalike physical parameters for which it is not straightforward to find a direct and universal, physical relationship. It is possible to stick to the empirical formulation in Equation 7.1 and make the slope S, the ground motion susceptibility, depend on the wind direction, respectively the friction velocity u_*, or the shear stress, in order to consider the influence of friction dynamically:

$$PSD_{rel} = S(u_*, site) \cdot u_{hor} + e_0(site). \tag{7.4}$$

In addition to the friction velocity, the polynomials S and e_0 still depend on specific site and instrument characteristics of the seismological station.

Considerations on the actual physical scaling between the applied force in the PBL and the PSD received confirm the empirical connection in Equation 7.4. Assuming frictional forces due to wind shear at the ground as main source for wind-induced seismicity, it is possible to estimate the load transmission. Emanating from the rate of energy applied at the ground, that must be equivalent to the work rate due to the shear, it is possible to express for a wind event of a finite source:

$$P_\tau = \frac{E_\tau}{t} \tag{7.5}$$

$$= \frac{\tau \cdot A \cdot r}{t}. \tag{7.6}$$

Thereby, P_τ is the power, respectively the rate of energy per time, applied at the surface. The energy due to shear can be expressed as shear force applied along a way r. Choosing a simple model of a storm with radial source with radius r and constant shear and mean wind speed, it is possible to write:

$$P = \tau \cdot \pi r^2 \cdot u, \tag{7.7}$$

depending on the mean wind speed u. Equation 7.7 thereby may give a rough idea of the applied power at the ground. Important is that the relation scales with the mean wind speed and the square of the friction velocity. Using the calculations of the turbulent meteorological parameters, estimating the spatial extent of a wind event, thus it is possible to derive the energy rate applied. The considerations however, built on the idea of deducing from one or multiple point measurements the mean wind field, and the turbulence of the wind field in the total area. Note, that turbulence finds itself in the microscale (Figure 2.3), that is rather not transferable to distances of more than tens of metres, depending on ground surface characteristics and the stratification.

The processes in the ground are as complex to estimate. From recordings of ground motion velocity, it is possible to calculate the PSD. The PSD thereby is not the total power, but the specific power in a given frequency band. The total power then would correspond to the average over all frequencies. As spectrograms of ground motion velocity in this study illustrate (i.e. Figure 6.15), the PSD is indeed a strongly frequency specific parameter. A comprehensive image of the power would need an average across all frequencies, of which seismological recordings provide a section, here: from 0.01 Hz to 100 Hz. Further, other than for meteorological in-situ measurements, the signals recorded at the seismological station, are not only the direct measure of signals at the station, but a superposition of signals from the complete source of excitation (Figure 6.13). Similar as assumed for the shear stress in the boundary layer, a most simple model would assume to have homogeneous and steady power received in the ground below the wind event. Unlike the meteorological in-situ measurements the seismological stations receive superpositioning signals from the surrounding environment as well, with lower impact, the longer the distance. Using the description of signals as wavefronts, propagating in the earth, they are subject to geometrical spreading (Stein and Wysession, 2009):

$$E(r) \propto E_0 \frac{1}{4\pi r^2}, \tag{7.8}$$

whereas $E(r)$ is the energy at a specific point of the spherical wavefront at distance r from the source. E_0 is the total energy radiated. The energy reaching a specific point obviously is a function of the distance from a point source. The same relation applies for the energy rate. Assuming the excitation of an extensive source at the surface, the wavefronts may be as long as the excitation source and the wavefronts parallel to the surface. The radius of the wavefronts then

$$r_s \rightarrow \infty. \tag{7.9}$$

The energy reaching at a specific point within the area of excitation then does not depend on the horizontal location of the receiver within the area of excitation. It is then possible to assume that the energy rate calculated at the seismological station is proportional to the applied energy rate:

$$P_{seis} \propto \tau \cdot \pi r^2 \cdot u. \tag{7.10}$$

Thereby, the power P_{seis} reaching a seismological station scales with the friction force, respectively the squared friction velocity, and the mean wind speed. Note, that Equation 7.10 represents a rough model. Nevertheless, it fits the empirical description in Equation 7.4 and supports the results of this section. However, building on many model assumptions here, the empirical, non-scaling description in Equation 7.4 is preferred at this point as significant finding.

7.2.2 Systematic Scaling of PSD of Ground Motion Velocity with Wind Direction

To interpret the distribution of the PSD of ground motion velocity with focus on wind direction, a polar interpolation of PSD data versus wind direction and wind speed is beneficial (calculations according to Chapter 5.3.4). Further, it allows to compare data from the whole seismological array qualitatively at once, to facilitate the interpretation in terms of the study area. Figure 7.11 shows the polar interpolation of PSD data with respect to wind direction and horizontal mean wind speed.

The white line marks the limit of data coverage and, similar to common meteorological wind roses, depicts the distribution of wind speed and wind direction during the study period (cf. Chapter 5.3.4). Strong winds above $10\,\mathrm{m\,s^{-1}}$ exclusively come from segments between west to south southwest, and southeast to east. From the northern hemisphere, the data base of wind speed above $5\,\mathrm{m\,s^{-1}}$ is sparse, and between the two major branches of high wind speed there is a gap of high wind speed data from south southeast. It is possible to draw profound conclusions from wind originating from the southern hemisphere.

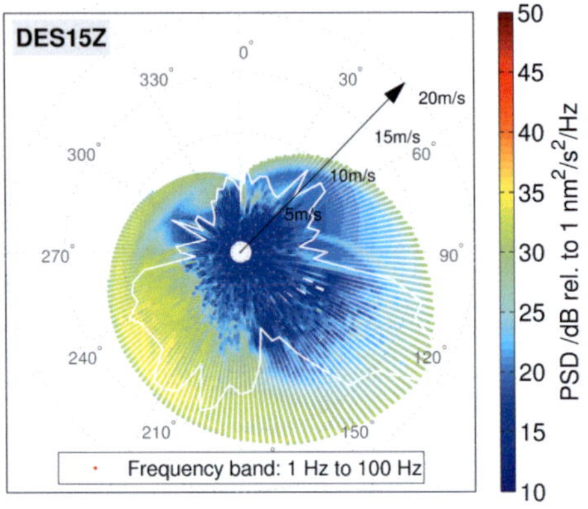

Figure 7.11: Polar interpolation of PSD of ground motion velocity with respect to wind speed and wind direction at Station DES15, Z-component from 01.11.2014 to 31.01.2015. The PSD is displayed as coloured dots with respect to the wind speed (radial axis) and the azimuth of the wind direction (polar axis). The white, solid line marks the area where data is available. Outside the line, points are interpolated without reference to existing data points. Frequency band of PSD data: 1 Hz and 100 Hz.

In essence, Figure 7.11 illustrates the context already discussed in Figure 7.10 (right panel) but the magnitude of the PSD of ground motion velocity is projected into a horizontal plane. The distribution of different affection of PSD to the inflow direction of the wind becomes visible. In all directions, the PSD increases with rising wind speed, respectively in outward direction. Passing the southern hemisphere from east to south towards west wind direction, the PSD rises for constant wind speed. Further, the vulnerability of the PSD to wind increases steadily. Winds from west have the strongest impact on ground motion velocity, winds from east the weakest. Even winds from south, running alongside the Dead Sea valley, have a stronger impact on ground motion velocity than the down slope winds from east.

The findings are systematic for all seismological stations of the array, even for Stations DES14 and GHAJ. Figure 7.12 shows a projection of polar descriptions of the PSD at all seismological DES Stations into the map of the study area (Stations DES14 and GHAJ separately in the Appendix, Figure A.5). It confirms the uniformity of the effect of wind with different inflow directions to all seismological stations. The PSD is shown with respect to the same colour

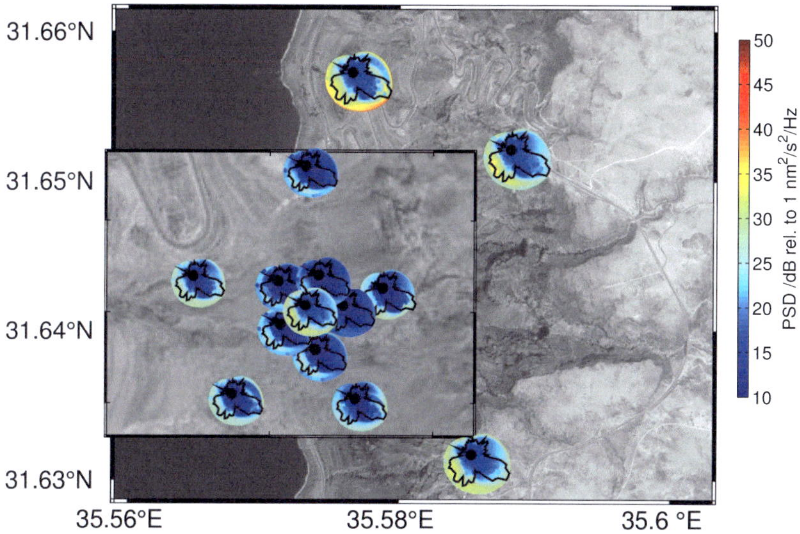

Figure 7.12: Polar interpolation of PSD at DES Stations DES01 to DES13 and DES15, with respect to wind speed and wind direction. Polar plots are located in the map at corresponding station locations (cf. Figure 4.2). Colours indicate the PSD with respect to the colour axis valid for all polar plots. Black lines in polar plots mark the data coverage, black dots mark the centres of the polar projections. Period: 01.11.2014 to 31.01.2015.

axis for all stations, therefore, the absolute colour level at some stations diverges from others. The relative effect at each station is consistent. The illustration of the whole array, revealing mostly uniform polar PSD distribution for all stations, independent of the location, confirms that the seismological stations are not dominantly influenced by in-situ wind. Instead, even in frequencies above 1 Hz they capture a more wide-ranging picture reaching beyond the area of the array. Equivalent calculations for July, 2014 with weaker winds, but better data coverage from northwest substantiate the results (Figure A.6).

Excitation of the Horizontal Ground Motion Velocity

The direction of the wind has a noticeable effect on the PSD of the ground motion velocity, according to Figure 7.12. Upslope wind flow enhances the PSD of the ground motion velocity stronger than wind from other directions. The question remains, whether wind from a specific direction polarizes the ground motion velocity, respectively its PSD. Horizontal particle motions of ground displacement during calm periods and in the presence of strong wind (Appendix, Figures A.9 and A.10) at the DES stations contradict the assumption that wind causes ground displacement dominantly polarised in the main wind direction. The particle motions show rather unsystematic patterns, individual at each seismological station and unrelated to the main wind direction.

The PSD of the horizontal ground motion velocity shows the same systematic behaviour as the PSD of the vertical ground motion velocity in Figure 7.12. However, the enhancement of the PSD of ground motion velocity in N- and E-component varies depending on the wind direction. The ratio of the PSD of N- and E-direction, depicted with respect to polar coordinates and indicating wind direction and wind speed (cf. in Chapter 5.3.4), illustrates the different magnitudes of enhancement of ground motion velocity in N- and E-direction. Figure 7.13 shows results for Stations DES01 and DES15 exemplary, Figure A.7 in the appendix depicts according results for the remaining seismological stations.

At station DES01 wind from southwest excites the PSD of ground motion velocity at the E-component stronger than the PSD at the N-component (blue areas). Wind from southeast and north excites the N-component of station DES01 stronger than the E-component (red areas). At station DES15, in contrast, the PSD of the ground motion velocity in E-direction is predominantly stronger than the PSD in N-direction. Whereas at both seismological stations wind from different directions systematically affects the power spectral density of the horizontal components in varying degree, the patterns extremely differ.

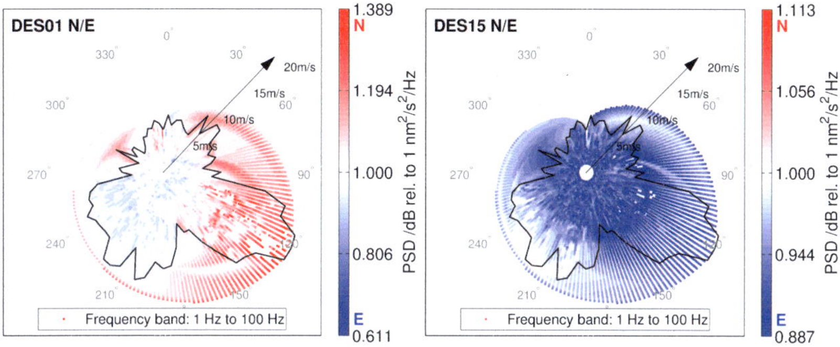

Figure 7.13: Polar interpolation of the relative PSD of ground motion velocity of the horizontal components N versus E with respect to wind speed and wind direction at stations DES01 and DES15 from 01.11.2014 to 31.01.2015. The ratio of the PSD of N-component versus E-component is displayed as coloured dots with respect to the wind speed (radial axis) and the azimuth of the wind direction (polar axis). The black, solid lines mark the area inside which data is available. Outside the line, points are interpolated without reference to existing data points. Frequency band: 1 Hz to 100 Hz.

Considering all seismological stations (Figure A.7) the pattern is systematic, but individual for each station. For a plain surface, one might expect the PSD to be strongest in the direction of the main wind direction during strong wind. Due to the complex topography in the study area, local effects, diverging the wind flow and hitting site-specific obstacles at the individual locations, might influence the direction of ground motion velocity excitement. The results reaffirm the ambient character of wind-induced microseisms (cf. Chapter 2.1.2).

7.2.3 Ground Motion Susceptibility with Respect to Topography

The estimation of the vulnerability of PSD of ground motion velocity to wind by calculating the ground motion susceptibility can be conducted analogue to the whole dataset in Section 7.1.2, Figure 7.3, but with the differentiation of multiple segments of wind direction. The table in Figure 7.14 summarizes the results for all seismological stations and components, distributed to 45° segments of wind direction. The dataset for each wind speed segment is smaller

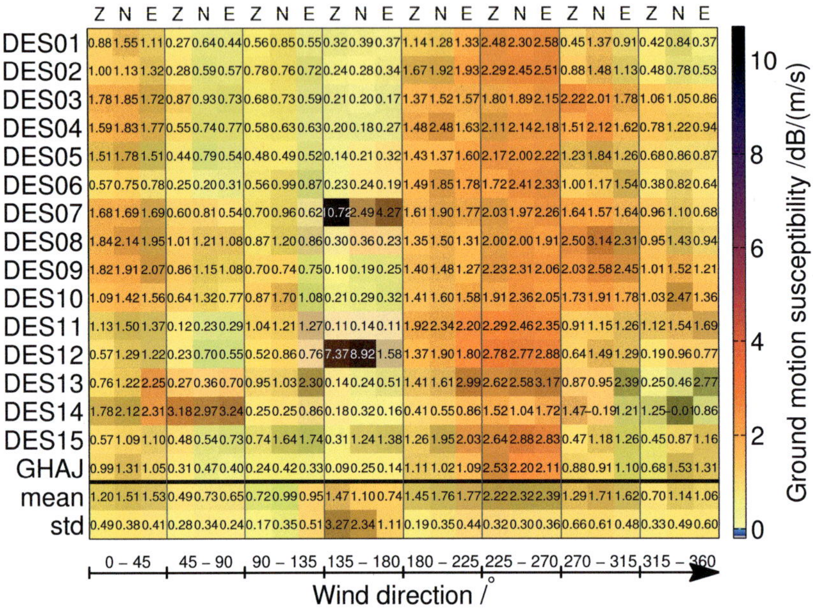

Figure 7.14: Ground motion susceptibility in $[S]$=1 dB m^{-1} s of PSD with wind speed for all stations depending on segments of wind direction. Frequency band: 1 Hz to 100 Hz. **Top to bottom:** seismological stations, mean, and standard deviation (std). Mean and standard deviation calculated for all stations except DES14 and GHAJ. **Left to right:** Wind directions in 45° segments. For each segment: Z, N, and E component of each station.

than the whole dataset, therefore, depending on the distribution of wind direction, the datasets are small compared to the full set used in Section 7.1.2 resulting in bigger errors and more outliers from the linear regression to the data points. This is especially the case for winds from 135° to 180°. The results are in accordance with the corresponding results for the whole dataset, less stable, but with the possibility to differentiate the flow direction of the wind.

Figure 7.15 illustrates the mean and standard deviations of the ground motion susceptibility to wind. The dependency on wind direction as derived from the

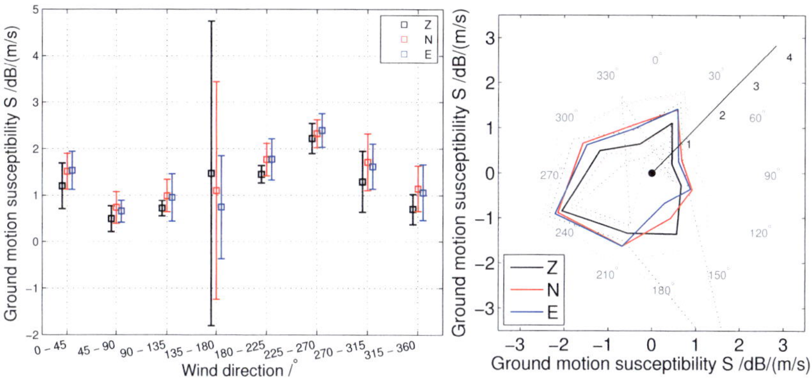

Figure 7.15: Mean of ground motion susceptibility of the DES Array for 45° segments of wind directions. Colours represent Z, N, and E components. **Left:** Susceptibility and standard deviation with respect to wind direction. **Right:** Polar representation of susceptibility (radial axis). Dotted lines represent standard deviations. Polar coordinates represent wind direction.

plane projections in Figure 7.12 can be confirmed by the susceptibility for the discrete wind directions, evident especially by the comparison with the polar illustration of the susceptibility. The susceptibility ranges from values of approximately $S = 0.5\,\mathrm{dB\,m^{-1}\,s}$ for winds from northeast and east southeast, to up to $S = 2.5\,\mathrm{dB\,m^{-1}\,s}$ for winds from west southwest, whereas the lowest susceptibility from northeast equals the susceptibility of the whole dataset confirming the results. The calculations are most stable for wind directions from west southwest and east southeast where the the datasets are most extensive and the wind speed is highest.

The results prove that the wind direction, respectively the topography, is an important factor when estimating the influence of wind on seismological records. Upslope winds towards higher elevations releasing higher stresses at the ground surface, have a significantly stronger impact on ground velocities than downslope winds or winds along the valley. It is possible to quantify the impact of wind on seismological records, and for plain terrain the more fundamental approach disregarding wind directions is probably sufficient.

7.3 Detection of Storms in Seismological Records

With the knowledge on the influence of wind on ground motion velocity the question arises, if the correlation of wind speed and seismological records can not only be measured quantitatively, but if wind can be detected solely from seismological records. Further, can the temporal course of wind speed be reproduced from seismological records?

The estimation of local wind speed from seismological records, if no local meteorological data should be available, would help to assess the wind-induced temporal change of seismological data quality. Similar as conducted in Chapter 6.3, comparing the course of wind speed and ground motion velocity, it is to calculate cross correlations of time dependent PSD data with the mean wind speed. Figure 7.16 shows the stacked sums of cross correlation coefficients for each seismological station and a wind speed record for the same period from MET1. The cross correlation coefficients are calculated for each day separately, whereas the sum of all daily cross correlation coefficients is divided by the total number of recording stations, as described in Chapter 5.2.3. The daily sum of cross correlation coefficients therefore can take values between -1 and 1.

The cross correlation coefficients reveal several incidents above 0.4, indicating minor, but definite similarities of the graphs. The periods often coincide with periods of increased wind speed above $10\,\mathrm{m\,s^{-1}}$. It is a strong indication that the PSD data run similar courses like the local wind speed.

The distribution of the high cross correlation coefficients leads to the assumption that, especially when taking advantage of the availability of multiple, similar seismological stations within one array, seismological records, can be used to reproduce the course of wind speed. Figure 7.17 compares the wind speed at MET1 from 01.11.2014 to 31.01.2015 with the temporal sequences of PSD

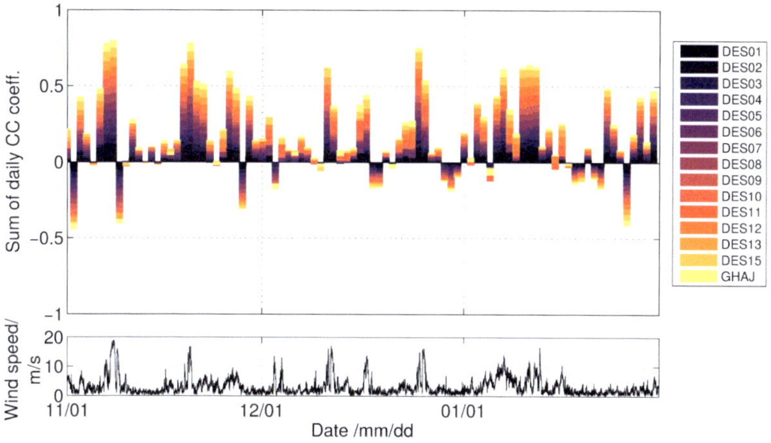

Figure 7.16: Daily cross correlation coefficients of PSD of ground motion velocity, Z-component, and mean wind speed at MET1 from 01.11.2014 to 31.01.2015. Coloured bars are stacked; the sum is divided by the number of recording stations. Frequency domain: 1 Hz to 100 Hz.

of ground motion velocity. The main graph shows the PSD tracks for all DES stations separately. All PSD tracks in Figure 7.17 are divided by an estimation of the PSD diurnal variations PSD_{dvar} (Figure A.8) to reduce the influence of anthropogenic noise, and therefore are noted without physical units. Calculating the average of all PSD tracks results in an estimation of the temporally dynamic, local noise field at the seismological stations. It is displayed in the bottom graph in comparison to the wind speed record at MET1.

The straightforward approach to stack all PSD traces and the reduction of the diurnal variations leads to a qualitative reproduction of the local wind field. The stacked PSD trace shows significant similarity with the mean wind speed, in fact, it maps incidents with wind speed above $5\,\mathrm{m\,s^{-1}}$ very detailed. So, without any presuppositions on the existing wind field in the calculations, it is possible, solely from seismological records of multiple stations, to maintain an estimation of the local wind field.

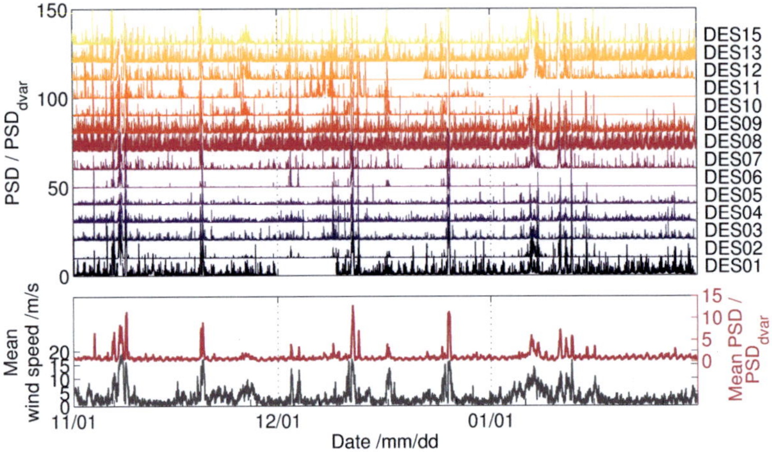

Figure 7.17: PSD traces, Z-component, divided by diurnal variations from 01.11.2014 to 31.01.2015. Frequency domain: 1 Hz to 100 Hz. **Top:** Separate PSD traces for the DES stations. **Bottom:** Mean PSD of all above traces (red) and mean wind speed at MET1 (grey).

7.4 Conclusions

The influence of wind on seismological records, investigated on the basis of an extensive, continuous dataset of multiple seismological stations and high frequent meteorological measurements, offers a broad range of possibilities to achieve the objectives. The examinations prove that time and frequency variant PSD of ground motion velocities are an ideal measure for the vulnerability of seismological records to wind.

Analogue to the empirical Moskowitz spectra that assessing the wave spectral density of ocean microseisms depending on various wind speed levels, spectra for seismological velocities with respect to wind speed levels are developed characterizing the wind dependency of seismological records. From the seismological wind spectra (Figure 7.2) it is possible to predict an estimation of the noise level of seismological records for specific wind speed levels. Figure 7.18 puts the wind dependent spectra in relation to NLNM and NHNM (cf.

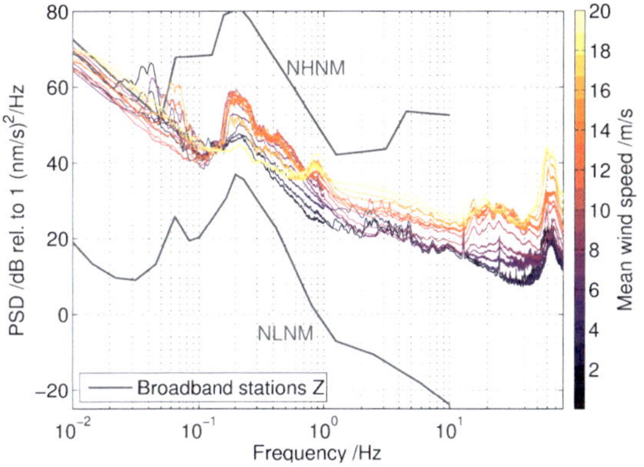

Figure 7.18: Noise level of the broad band stations of the DES Array from 01.11.2014 to 31.01.2015 depending on the mean wind speed. For comparison, the NLNM and the NHNM for ground motion velocity (grey lines) are given additionally. NLNM and NHNM data from: Bormann (1998)

Figure 2.2). Wind excites the PSD of seismic ground motion velocity apparently. Wind speed of $20\,\mathrm{m\,s^{-1}}$ lifts the PSD level by about a quarter of the gap between NLNM and NHNM.

Further, by defining the ground motion susceptibility it is possible to obtain a general relation for the vulnerability of ground motion velocity and wind speed. Although the friction velocity (Figure 7.9) is a more physically adequate parameter to estimate the shear on the earth's surface, and more directly related to the original triggering mechanism for wind induced seismic signals, instead the horizontal wind speed is used as a measure to assess the magnitude of wind related signals. It turns out, that the mean horizontal wind speed is a fairly adequate parameter to evaluate the wind induced stress affecting the ground movement, at least up to wind speeds of about $20\,\mathrm{m\,s^{-1}}$ obtained in this study.

Section 7.3 introduces a study to demonstrate the magnitude of correlation of wind speed records and PSD of ground motion velocity. It is possible to read the change of local wind speed from the seismological records above 1 Hz and to detect the presence of a storm without a priori information from meteorological measurements.

8 Summary and Discussion

In this study diverse approaches are developed to investigate the influence of the dynamic processes in the atmosphere, respectively the surface layer, on ground motion velocity. Thereby, several key questions are addressed:

i Which processes at the interface between atmosphere and solid earth interact, causing ambient vibrations at the earth's surface? How do applied forces scale and manifest in seismological recordings?

ii Ground motion susceptibility S: does a scaling between commonly used meteorological and seismological parameters exist, describing or representing the actual physical processes? Is there a systematic scaling between wind speed and seismological parameters? And down to which level does wind cause measurable ground motions?

iii Following the idea of spectra describing noise in seismological records: is it convenient to develop a wind speed-related noise model for ground motion velocity? Which frequencies are affected by the influence of wind, following the long-term exposure of recordings to wind?

iv Considering the signal sources: besides the local wind field, how do severe wind events of larger extent affect seismological records?

Since there exist only few systematic studies on referred topics, the research questions are fundamental and manifold. To answer the questions, likewise various approaches are adopted.

Using an elaborate dataset of ground motion velocity records from 15 temporary seismological stations and wind speed records from one meteorological station, assessed over the same period of one year at the Dead Sea, provides the opportunity to establish one common MS dataset, convenient for systematic examinations. Adding observations from a permanent seismological station as a reference shows that the examined influence of wind on seismological records is not the cause of insufficient data quality of the temporary stations nor is it a consequence of the near-surface measurements. Therefore, results of this study can be expected to have transferable character.

8.1 Empirical Findings and Theoretical Implications

To investigate the impact of wind on ground motion velocity the consideration of horizontal wind speed as indicatory parameter to assess the impact of shear on the earth's surface seems most likely. From the physical point of view not the wind speed itself, but the wind shear, acting above the solid surface, is crucial. Other than the friction velocity the wind speed only indirectly scales with the shear stress at the discontinuity between solid earth and atmosphere. As a first approximation surface characteristics, such as roughness length, topography, or the presence of obstacles, are disregarded. The parameter of the horizontal wind speed roughly scales with the friction velocity, respectively with the shear stress, and therefore is found to be sufficient to conduct systematic examinations as a first attempt.

Ground Motion Susceptibility

As wind-related influence on seismological records can be characterised as a noise source with predominantly ambient character, it is described in form of PSD instead of ground motion velocity time series. Results from the Dead Sea area show that the PSD of the ground velocities, collocated with local wind speed records, rise with increasing wind speed. The ground motion susceptibility S, defined in this study as a measure of the lowest PSD level of ground motion velocity with respect to the horizontal wind speed, strongly depends on the frequency range. The average ground motion susceptibility for the seismological array is:

- 0.01 Hz - 0.1 Hz: $S > 1$ dB m^{-1} s
- 0.1 Hz - 01 Hz: $S \approx 0.5$ dB m^{-1} s
- 1 Hz - 100 Hz: $0.7 < S < 1$ dB m^{-1} s

Like the background noise level, the ground motion susceptibility is higher for the horizontal recording components than for the vertical component throughout the dataset (Figure 7.6). The definition of the ground motion susceptibility is a rather cautious approximation. It marks the lower boundary of the PSD of ground motion velocity, represented by the 5% percentile, waiving diversification of the PSD towards higher levels with increasing wind speed. The lower limit of the PSD thereby already rises in calm wind regimes where the mean wind speed does not exceed 2 m s^{-1}.

The results do not only confirm that the presence of wind has a systematic effect on ground motion velocity, respectively its PSD. The impact actually is striking throughout the whole frequency band of seismological records: the PSD of the ground motion velocity increases by about 1 dB with each 1 m s^{-1} increase of horizontal wind speed. Considering the offset range of 20 dB to 70 dB between NLNM and NHNM (cf. Figure 2.2), the ground motion susceptibility of 1 dB m^{-1} s indicates that wind has a significant effect on the data quality. A fresh breeze, wind speed of up to 10 m s^{-1}, enhances the PSD level by up to 10 dB compared to a calm wind regime.

In this study, the ground motion susceptibility is developed as a linear factor between the logarithmic PSD level and the wind speed (cf. Equation 7.1). The linear dependency is confirmed by a case study from Dietze et al. (2015). In a different context, investigating the influence of neighbouring wind farms to ground motion velocity measurements, Saccorotti et al. (2011) find a power law scaling of PSD with wind speed. However, ignoring wind speed regimes below 2 m s^{-1}, where results tend to be insignificant, the relation found by Saccorotti et al. (2011) could be linear as well. The results allow the interpretation of both a linear scaling and a power scaling of the PSD level without noteworthy deviations (Figure 7.3). The factual relation must be expected to be much more complex than both approximations (cf. model assumptions on seismic noise in Chapter 2.1.2), as a lot of parameters, such as the turbulence, the friction, site specific obstacles, and the dimensions of the wind regime are not considered here. Using local wind speed measurements as only reference on the meteorological situation, is a strong, but sufficient simplification, therefore the interpretation of the results is kept as elementary in this study, remaining with the model of the ground motion susceptibility.

Wind Speed Related Spectra

Establishing the station specific ground motion susceptibility is a feasible way to predict the approximate noise level for certain wind regimes based on the PSD level. More general, but frequency specific estimations on the effect of wind on the noise level of seismological records can be assessed by wind speed depending PSD spectra (Figure 7.2). From the wind speed depending spectra, it is possible to estimate the PSD level for specific wind speed levels, similar as to classify the seismological data quality according to the NLNM. The empirical wind speed related spectra, developed in this study, illustrate the vulnerability of ground motion velocity records with respect to specific frequencies. At one glance it appears that the presence of wind, even light air, has an influence of the noise level in the complete frequency band, the strongest impact for frequencies above 1 Hz.

While the data base from seismological stations solely within the Dead Sea region and for the period of three months may be sufficient for a model for the Dead Sea region, it may be too sparse and partial to derive a universal model. The author of this work proposes the equivalent creation of a more general, world wide model, based on long periods of high-quality seismological records from permanent stations with access to local wind measurements.

Classification of Wind-Induced Signal Sources

The identification and classification of wind-induced signals, based on case studies, does not contribute to analyse the wind-related level of seismicity in seismological records. However, case studies on specific wind events help to learn about how regional characteristics or meteorological events with specific characteristics influence the ground motion velocity.

The case studies in Chapter 6 confirm results from previous studies (i.e. Dahm et al., 2006), identifying the main impact of microseisms between 0.08 Hz and 0.2 Hz originating from primary and secondary ocean waves. In this study, they are assumed to be linked to the Mediterranean Sea (Figure 6.22). Further, data in this study reveal distinct signals in the frequency domain between 0.4 Hz and 1 Hz temporally correlating with local wind speed measurements, particularly with wind speed records from Station EBS3 close to the Dead Sea shore. Confirmed by studies on microseisms assigned to lakes (Lynch, 1952), it can be assumed that the signals may be related to water waves from the Dead Sea. Other than the uniform signals below 1 Hz, signals in frequencies above 1 Hz can be assigned to local wind, showing variational character, individual for all stations.

The investigation of the case studies on March 2014 and January 2015 clarifies that wind has very individual effect on seismological records. Depending on the region, the presence of obstacles, lakes, and on the genesis and extent of storms, different frequency ranges of ground motion velocity are excited. Signals that can be assigned to ocean waves, agitated by wind, can travel long distances and show uniform character at all regionally distributed stations. In contrast, signals associated to local wind, act very site specific (Figure 6.15). The results show that the influence of wind on seismological records is hard to unify. By averaging data from the whole array and by deriving systematic dependencies of PSD levels to wind speed over a long period, however, general approximations can be assessed. Therefore, it is even more important to learn about wind-induced noise.

While wind-induced signals in seismological records in frequencies below 1 Hz can be assigned to microseisms from the ocean or the Dead Sea by the signal characteristics, there is no close correlation with the course of the local wind speed. The local wind speed becomes strongly apparent in seismological records in the high frequencies, confirmed in case studies on severe wind events (Chapters 6.3.1 and 7.3) by high cross-correlation coefficients of up to 0.85 between wind speed and power time series of ground motion velocity. It is even possible to reproduce the approximate course of the local wind speed from the power time series, averaged for the whole seismological array (Figure 7.17). Stacking power time series of multiple close seismological stations, it is possible to detect storms without a priori knowledge from meteorological data.

8.2 Deriving Applied Forces from Seismological Records

The scaling of the PSD of the ground motion velocity with the horizontal wind speed is a feasible approach. However, the approximation does not reproduce the extensive, real physical processes originating from shear-induced friction at the surface. The strong topography of the Dead Sea valley enables to investigate the influence of different frictional regimes, differentiated by wind from specific directions. The friction velocity (Equation 2.6), and resulting the shear stress, increases as expected with rising wind speed. Asperities at the surface, assigned to topography, change the released stress due to air flow, presumably adding pressure forces to the friction forces (Figure 8.1).

Results from the Dead Sea region confirm the considerations on wind flow (Figure 8.1), exhibiting specific wind shear characteristics for the main wind directions along and across the valley. Up slope winds in the Dead Sea valley are subject to higher wind shear than valley winds along the valley. Down slope wind coincides with the overall lowest shear (Figure 7.9). The results

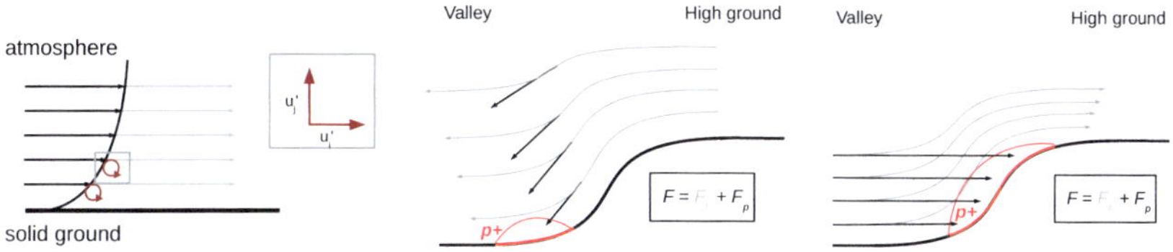

Figure 8.1: Applied forces between solid ground and atmosphere. Grey arrows indicate wind trajectories. F_f: friction forces, F_p: dynamic pressure forces. **Left:** even terrain. **Middle:** elevated terrain, down slope wind. **Left:** elevated terrain, up slope wind.

on shear in the atmosphere match the findings on ground motion velocity. A similar dependency on wind direction can be determined from the PSD of the ground motion velocity as well. It suggests that wind shear in the surface layer, due to friction, or indirectly by additional dynamic pressure towards obstacles, excites ground motion velocity. Thereby, rugged terrain enhances the impact of wind on seismological records noticeably and systematically (Chapter 7.2). The ground motion susceptibility fluctuates accordingly between

$$0.5\,\mathrm{dB\,m^{-1}\,s} < S < 2.2\,\mathrm{dB\,m^{-1}\,s}.$$

The lowest values of the ground motion susceptibility occur for down slope winds from east, the highest impact is generated for up slope winds from south-west, altering the ground motion susceptibility by a factor of more than 4. The intense variations imply that the topography plays a major role when observing the wind-induced vibration of the earth's surface.

It is realistic to anticipate that the stress release in the surface layer, estimated by the friction velocity, roughly corresponds to stress applied at the solid ground, causing ground vibrations. Deriving shear stress from meteorological turbulent measurements may lead to the forces released in the ground. Theoretical concepts already are ahead. De Langre (2008), for example, pursue a manifold approach, modelling forces released by wind acting on diverse surfaces, also including vegetation. Theoretically, forces in the ground, related

to wind, can be determined by wind shear. In practice, however, it is not yet possible to deduce how the corresponding forces would appear in seismological records at one specific station. As outlined in Chapter 2.1.2, wind is a rather ambient source, complexly distributed over extensive surfaces. Therefore, ground motion velocity at a specific receiver can not yet be used to derive the original stress or forces, as it can be modelled as source parameters for tectonic ruptures.
According to this study, for the present, the research remains with empirical scaling approaches of energy densities of ground motion velocity PSD_{seis}. It depends on the mean horizontal wind speed u_{hor}, representing a simplification of the 3D wind field, and more complex, if available, on the friction velocity u_*:

$$PSD_{seis} \vdash u_{hor}, u_*, C_{met}, C_{surf}. \tag{8.1}$$

C_{met} and C_{surf} indicate meteorological and surface characteristics, such as the spatial extent of the wind source, the distance to the receiver, topography, and obstacles, which may presumably be parametrised in future studies.

8.3 Conclusions

This study takes its place alongside few existing studies on wind-induced seismicity. While some studies investigate the influence of wind using case studies on storms, handling the results as wind-induced signals, others approach the subject in terms of an overall noise level, related to wind. In contrast, the current study follows a multi-pronged approach. Case studies, essential to verify and assign signals that are characteristic for the Dead Sea region, and long, continuous periods, comprising an extensive data base, lead to a more thorough understanding of wind-induced seismicity.
The fundamental character of the raised questions requires manifold approaches, all based on the design of a MS dataset, essentially building on seismological spectrogram data, as introduced in this study. Thereby, PSD data of ground

motion velocity proves to be a convenient and applicative parameter to display wind-induced signals in the form of spectrograms, allowing to establish a systematic dependency of ground motion velocity with respect to the wind speed. The development of the ground motion susceptibility as parameter to describe the vulnerability of seismological records to wind is a transferable parameter, proposed to apply in future studies as well.
Results confirm that wind has a striking impact on seismological records, not only inducing high amplitudes, but at the same time affecting a broad frequency band. The wind speed related spectra, covering three months of continuous recordings at the seismological array are a valid illustration of the empirical findings.

Bibliography

Aki, K. (1967). Scaling law of seismic spectrum. *J. Geophys. Res.*, 72(4):1217–1231.

Aki, K. and Richards, P. (1980). *Quantitative Seismology*. Freeman, San Francisco, 1 edition. Vol. 1: 557 pp., Vol. 2: 373 pp.

Al-Eisawi, D. (1996). *Vegetation of Jordan*. Regional Office for Science and Technology for the Arab States. UNESCO, Cairo. viii. 284 pp.

Almendros, J., Ibáñez, J. M., Alguacil, G., and Del Pezzo, E. (1999). Array analysis using circular-wave-front geometry: an application to locate the nearby seismo-volcanic source. *Geophys. J. Int.*, 136(1):159–170.

Alpert, P., Cohen, A., Neumann, J., and Doron, E. (1982). A model simulation of the summer circulation from the eastern mediterranean past Lake Kinneret in the Jordan Valley. *Mon. Weather Rev.*, 110(8):994–1006.

Alpert, P., Neeman, B., and Shay-El, Y. (1990). Intermonthly variability of cyclone tracks in the Mediterranean. *J. Climate*, 3(12):1474–1478.

Alpert, P., Osetinsky, I., Ziv, B., and Shafir, H. (2004a). A new seasons definition based on classified daily synoptic systems: an example for the eastern Mediterranean. *Int. J. Climatol.*, 24(8):1013–1021.

Alpert, P., Osetinsky, I., Ziv, B., and Shafir, H. (2004b). Semi-objective classification for daily synoptic systems: Application to the eastern Mediterranean climate change. *Int. J. Climatol.*, 24:1001–1011. doi: 10.1002/joc.1036.

Asch, G. (2009). Seismic recording systems. in: Bormann, P. (ed.), new manual of seismological observatory practice (NMSOP). *GFZ German Research Centre for Geosciences, Potsdam; http://nmsop.gfz-potsdam.de*, pages 1–20. doi: 10.2312/GFZ.NMSOP_r1_ch6.

Beauduin, R., Lognonné, P., Montagner, J., Cacho, S., Karczewski, J., and Morand, M. (1996). The effects of the atmospheric pressure changes on seismic signals or how to improve the quality of a station. *Bull. Seismol. Soc. Am.*, 86(6):1760–1769.

Bendat, J. S. and Piersol, A. G. (2010). *Random data: analysis and measurement procedures.* Wiley series in probability and statistics. John Wiley & Sons, 4th edition. xxi, 604 pp.

Bonnefoy-Claudet, S., Cotton, F., and Bard, P.-Y. (2006). The nature of noise wavefield and its applications for site effects studies: A literature review. *Earth-Science Reviews*, 79(3-4):205 – 227.

Bormann, P. (1998). Conversion and comparability of data presentations on seismic background noise. *J. Seismol.*, 2(1):37–45.

Bormann, P. E. (2012). New manual of seismological observatory practice (NMSOP-2), IASPEI. *GFZ German Research Centre for Geosciences, Potsdam; http://nmsop.gfz-potsdam.de.* Ed. 2, doi: 10.2312/GFZ.NMSOP-2.

Brandt, S. and Dahmen, H. (2005). *Mechanik-Eine Einführung in Experiment und Theorie Springer-Verlag.* Berlin, 4th edition. xiii, 561 pp.

Bromirski, P. D. and Duennebier, F. K. (2002). The near-coastal microseism spectrum: Spatial and temporal wave climate relationships. *J. Geophys. Res. B: Solid Earth*, 107(B8):ESE 5–1–ESE 5–20.

Bromirski, P. D., Duennebier, F. K., and Stephen, R. A. (2005). Mid-ocean microseisms. *Geochem. Geophy. Geosy.*, 6(4):1–19.

Brune, J. (1970). Tectonic stress and the spectra of seismic shear waves from earthquakes. *J. Geophys. Res.*, 75(26):4997–5009.

Capon, J. (1973). Signal processing and frequency-wavenumber spectrum analysis for a large aperture seismic array. *Methods in Computational Physics*, 13:1–59.

Carlson, J. and Foster, M. (1986). Numerical study of some neutrally and unstably stratified boundary-layer flows over a valley at small Richardson number. *Tellus A*, 38(1):60–75.

Castellaro, S. and Mulargia, F. (2012). A statistical low noise model of the Earth. *Seismol. Res. Lett.*, 83(1):39–48.

Cécé, R., Bernard, D., d'Alexis, C., and Dorville, J.-F. (2014). Numerical simulations of island-induced circulations and windward katabatic flow over the Guadeloupe archipelago. *Mon. Weather Rev.*, 142(2):850–867.

Cessaro, R. K. (1994). Sources of primary and secondary microseisms. *Bull. Seismol. Soc. Am.*, 84(1):142–148.

Dahm, T., Krüger, F., Essen, H.-H., and Hensch, M. (2005). Historic micro-seismic data and their relation to the wave-climate in the North Atlantic. *Meteorol. Z.*, 14(6):771–779.

Dahm, T., Tilmann, F., and Morgan, J. (2006). Seismic broadband ocean-bottom data and noise observed with free-fall stations: Experiences from long-term deployments in the North Atlantic and the Tyrrhenian Sea. *Bull. Seismol. Soc. Am.*, 96(2):647–664.

Davison, C. (1924). Prof. Fusakichi Omori. *Nature*, 113:133.

De Karman, T. and Howarth, L. (1938). On the statistical theory of isotropic turbulence. *Proc. R. Soc. A*, 164:192–215.

De Langre, E. (2008). Effects of wind on plants. *Annu. Rev. Fluid Mech.*, 40:141–168.

Dewey, J. and Byerly, P. (1969). The early history of seismometry (to 1900). *Bull. Seismol. Soc. Am.*, 59(1):183–227.

Dietze, M., Iserloh, T., Seeger, M., Simard, S., Turowski, J., and Hovius, N. (2015). Using seismometers to study modes and intensities of rainfall events. In *AK Geomorpholigie Berlin, Oct. 03-04*. Conference contribution.

Draganov, D., Wapenaar, K., Mulder, W., Singer, J., and Verdel, A. (2007). Retrieval of reflections from seismic background-noise measurements. *Geophys. Res. Lett.*, 34(L04305):1–4.

Durran, D. R. (1990). Mountain waves and downslope winds. *Meteor. Monogr.*, 23(45):59–81.

Edwards, W. N. and Green, D. N. (2012). Effect of interarray elevation differences on infrasound beamforming. *Geophys. J. Int.*, 190(1):335–346.

Egger, J. (1990). Thermally forced flows: Theory. *Atmospheric Co-Processes over Complex Terrain, Meteor. Monogr.*, (45):43–57.

Essen, H.-H., Krüger, F., Dahm, T., and Grevemeyer, I. (2003). On the generation of secondary microseisms observed in northern and central Europe. *J. Geophys. Res. B: Solid Earth*, 108(B10):ESE 15.

Etling, D. (2008). *Theoretische Meteorologie*, volume 3. Springer, 3., erw. und aktual. aufl edition. Chapter 18, 19, 359 pp.

Foken, T. (2006). 50 years of the monin–Obukhov similarity theory. *Bound.-Lay. Meteorol.*, 119(3):431–447.

Friedrich, A., Krüger, F., and Klinge, K. (1998). Ocean-generated microseismic noise located with the Gräfenberg array. *J. Seismol.*, 2(1):47–64.

GEOFON Data Centre (1993). GEOFON seismic network. *Deutsches Geo-ForschungsZentrum GFZ*. doi: 10.14470/TR560404.

Gerstoft, P., Shearer, P. M., Harmon, N., and Zhang, J. (2008). Global P, PP, and PKP wave microseisms observed from distant storms. *Geophys. Res. Lett.*, 35(23):1–6. L23306.

Gesterman, N., B., S., P., L., and M., J. (2015). Detection capability estimation of the IFE14 SAMS network. *Science and Technology Conference, CTBTO Vienna*.

Groos, J. (2007). Analyse des seismischen Rauschens im Großraum Bukarest. Diploma thesis, Universität Karlsruhe.

Groos, J. (2010). *Broadband Seismic Noise : Classification and Greens Function Estimation*. PhD thesis, Karlsruhe Institute of Technology.

Groos, J. C. and Ritter, J. R. R. (2009). Time domain classification and quantification of seismic noise in an urban environment. *Geophys. J. Int.*, 179(2):1213–1231.

Hall, J. K. (1996). Digital topography and bathymetry of the area of the Dead Sea depression. *Tectonophysics*, 266(1):177–185.

Harjes, H. and Henger, M. (1973). Array-seismologie. *Zeitschrift für Geophysik*, 39:865–905.

Haskell, N. (1964). Total energy and energy spectral density of elastic wave radiation from propagating faults. *Bull. Seismol. Soc. Am.*, 54(6A):1811–1841.

Herzog, A. (2015). Analyse lokaler Windsysteme am Toten Meer mittels der Daten des DESERVE Messmastes. B.S. Thesis, Karlsruhe Institute of Technology.

Holub, K., Rušajová, J., and Sandev, M. (2008). The January 2007 windstorm and its impact on microseisms observed in the Czech Republic. *Meteorol. Z.*, 17(1):47–53.

Holub, K., Rušajová, J., and Sandev, M. (2009). A comparison of the features of windstorms Kyrill and Emma based on seismological and meteorological observations. *Meteorol. Z.*, 18(6):607–614.

Hu, H. and Wei, N. (2009). A study of GPS jamming and anti-jamming. *2009 2nd International Conference on Power Electronics and Intelligent Transportation System*, pages 388–391. Conference Paper.

Jarvis, A., Reuter, H., Nelson, A., and Guevara, E. (2008). Hole-filled seamless SRTM data V4. Available from: http://srtm.csi.cgiar.org.

Kafle, H. K. and Bruins, H. J. (2009). Climatic trends in Israel 1970–2002: warmer and increasing aridity inland. *Clim. Change*, 96(1-2):63–77.

Kalthoff, N. and Vogel, B. (1992). Counter-current and channelling effect under stable stratification in the area of Karlsruhe. *Theor. Appl. Climatol.*, 45(2):113–126.

Kibblewhite, A. C. and Ewans, K. C. (1985). Wave-wave interactions, microseisms, and infrasonic ambient noise in the ocean. *J. Acoust. Soc. Am.*, 78(3):981–994.

Koletsis, I., Lagouvardos, K., Kotroni, V., and Bartzokas, A. (2009). The interaction of northern wind flow with the complex topography of Crete Island– Part 1: Observational study. *Nat. Hazards Earth Syst. Sci.*, 9(6):1845–1855.

Kottmeier, C., Agnon, A., Al-Halbouni, D., Alpert, P., Corsmeier, U., Dahm, T., Eshel, A., Geyer, S., Haas, M., Holohan, E., Kalthoff, N., Kishcha, P., Krawczyk, C., Lati, J., Laronne, J. B., Lott, F., Mallast, U., Merz, R., Metzger, J., Mohsen, A., Morin, E., Nied, M., Rödiger, T., Salameh, E.,

Sawarieh, A., Shannak, B., Siebert, C., and Weber, M. (2016). New perspectives on interdisciplinary earth science at the Dead Sea: The DESERVE project. *Sci. Total Environ.*, 544:1045–1058.

Kraus, H. (2004). *Die Atmosphäre der Erde: Eine Einführung in die Meteorologie*. Springer-Verlag, 3rd edition. xv, 422 pp.

Krumgalz, B. S., Hecht, A., Starinsky, A., and Katz, A. (2000). Thermodynamic constraints on Dead Sea evaporation: can the Dead Sea dry up? *Chem. Geol.*, 165(1):1–11.

Lay, T. and Wallace, T. C. (1995). *Modern global seismology*. International geophysics series; 58. Academic Press. xii, 521 pp.

Lee, X., Finnigan, J., et al. (2004). Coordinate systems and flux bias error. In *Handbook of Micrometeorology*, pages 33–66. Springer.

Lensky, N., Dvorkin, Y., Lyakhovsky, V., Gertman, I., and Gavrieli, I. (2005). Water, salt, and energy balances of the Dead Sea. *Water Resour. Res.*, 41(12):1–13. W12418.

Lesouëf, D., Gheusi, F., Delmas, R., and Escobar, J. (2011). Numerical simulations of local circulations and pollution transport over Reunion island. *Ann. Geophys*, 29:53–69.

Longuet-Higgins, M. (1950). A theory of the origin of microseisms. *Phil. Trans. R. Soc. Lond. A*, 243(857):1–35.

Lott, F., Al-Qaryouti, M., Corsmeier, U., and Ritter, J. (2016). Dead sea seismic array, Jordan for DESERVE project (Feb. 2014 - Feb. 2015). *Scientific Technical Report STR16/01*, 16(01):1–11. doi: http://doi.org/10.2312/GFZ.b103-16011.

Lynch, J. (1952). The Great Lakes, a source of two-second frontal microseisms. *Eos, Trans. Amer. Geophys. Union*, 33(3):432–434.

Masters, T. and Widmer, R. (1995). Free oscillations: Frequencies and attenuations. *Global Earth Physics: a handbook of physical constants*, pages 104–125.

Matsuzawa, T., Obara, K., Maeda, T., Asano, Y., and Saito, T. (2012). Love- and Rayleigh-wave microseisms excited by migrating ocean swells in the North Atlantic detected in Japan and Germany. *Bull. Seismol. Soc. Am.*, 102(4):1864–1871.

McNamara, D., Hutt, C., Gee, L., Benz, H. M., and Buland, R. (2009). A method to establish seismic noise baselines for automated station assessment. *Seismol. Res. Lett.*, 80(4):628–637.

McNamara, D. E. and Buland, R. P. (2004). Ambient noise levels in the continental United States. *Bull. Seismol. Soc. Am.*, 94(4):1517–1527.

Metzger, J., Nied, M., Corsmeier, U., Kleffmann, J., and Kottmeier, C. (submitted 2016). Energy balance, evaporation rates and dew deposition of the Dead Sea and its surroundings. *Bound.-Lay. Meteorol.*

Moeng, C.-H. and Sullivan, P. P. (1994). A comparison of shear-and buoyancy-driven planetary boundary layer flows. *J. Atmos. Sci.*, 51(7):999–1022.

Mölders, N. and Kramm, G. (2014). *Lectures in Meteorology*. Springer. xix, 591 pp.

Moskowitz, L. (1964). Estimates of the power spectrums for fully developed seas for wind speeds of 20 to 40 knots. *J. Geophys. Res.*, 69(24):5161–5179.

Mucciarelli, M., Gallipoli, M. R., Di Giacomo, D., Di Nota, F., and Nino, E. (2005). The influence of wind on measurements of seismic noise. *Geophys. J. Int.*, 161(2):303–308. doi: 10.1111/j.1365-246X.2004.02561.x.

Nadeau, R. and Dolenc, D. (2005). Nonvolcanic tremors deep beneath the San Andreas fault. *Science*, 307(5708):389.

NSTB (2014). *Global Positioning System (GPS) Standard Positioning System (SPS) Performance Analysis Report*, volume Report #89. NSTB WAAS Atlantic City International Airport, NJ 08405; Federal Aviation Administration, William J.Hughes Technical Center, Atlantic City, NJ, April 30. 62 pp.

Omori, F. (1899). Horizontal pendulums for the mechanical registration of seismic and other earth movements. *Tokyo Imperial University*.

Orlanski, I. (1975). A rational subdivision of scales for atmospheric processes. *Bull. Am. Meteorol.Soc.*, 56:527–530.

Percival, D. B. and Walden, A. T. (1993). *Spectral analysis for physical applications*. Cambridge University Press, 1st edition. xxvii, 583 pp.

Peterson, J. (1993). Observations and modeling of seismic background noise. *Open file report*, 93-322.

Pierson, W. and Moskowitz, L. (1964). A proposed spectral form for fully developed wind seas based on the similarity theory of S.A. Kitaigorodski. *J. Geophys. Res.*, 69(24):5181–5190.

Raupach, M. and Finnigan, J. (1997). The influence of topography on meteorogical variables and surface-atmosphere interactions. *J. Hydrol.*, 190(3):182–213.

Richter, C. F. (1935). An instrumental earthquake magnitude scale. *Bull. Seismol. Soc. Am.*, 25(1):1–32.

Ritter, J. and Groos, J. (2007). Kyrills seismischer Fingerabdruck. *Spektrum Wiss.*, page 19.

Rost, S. and Thomas, C. (2002). Array seismology: methods and applications. *Rev. Geophys.*, 40(3):2–1–2–27.

Ryberg, T., Weber, M. H., Garfunkel, Z., and Bartov, Y. (2007). The shallow velocity structure across the Dead Sea Transform fault, Arava Valley, from seismic data. *J. Geophys. Res. B: Solid Earth*, 112(B8):1–12. B08307.

Saaroni, H., Ziv, B., Bitan, A., and Alpert, P. (1998). Easterly wind storms over Israel. *Theor. Appl. Climatol.*, 59(1):61–77.

Saccorotti, G., Piccinini, D., Cauchie, L., and Fiori, I. (2011). Seismic noise by wind farms: a case study from the Virgo Gravitational Wave Observatory, Italy. *Bull. Seismol. Soc. Am.*, 101(2):568–578.

Sahmoudi, M. and Moeness, G. (2009). Robust tracking of weak GPS signals in multipath and jamming environments. *Signal Process.*, 89:1320–1333. doi:10.1016/j.sigpro.2009.01.001.

Scherbaum, F. and Schmidtke, E. (2007). *Of poles and zeros : fundamentals of digital seismology*. Modern approaches in geophysics ; 15. Springer, Dordrecht, 2. ed., repr. with corr. edition. 268 pp.

Schulte-Pelkum, V., Earle, P. S., and Vernon, F. L. (2004). Strong directivity of ocean-generated seismic noise. *Geochem. Geophy. Geosy.*, 5(3):1–13.

Schweitzer, J., Fyen, J., Mykkeltveit, S., Gibbons, S. J., Pirli, M., Kühn, D., and Kvarna, T. (2012). Seismic arrays. in: Bormann, p. (ed.), new manual of seismological observatory practice 2 (nmsop-2). *GFZ German Research Centre for Geosciences, Potsdam; http://nmsop.gfz-potsdam.de*, pages 1–80. doi: http://doi.org/10.2312/GFZ.NMSOP-2_ch9.

Shafir, H., Jin, F., Lati, Y., Cohen, M., and Alpert, P. (2008). Wind channeling by the Dead-Sea Wadis. *Open Atmospheric Science Journal*, 2:139–152.

Shapiro, N. M. and Campillo, M. (2004). Emergence of broadband Rayleigh waves from correlations of the ambient seismic noise. *Geophys. Res. Lett.*, 31(7):1–4. L07614.

Sharp, J. M. (2002). Columbia Gorge gap flow. *Bull. Am. Meteorol. Soc.*, 83(12):1757.

Shelly, D. R., Beroza, G. C., and Ide, S. (2007). Non-volcanic tremor and low-frequency earthquake swarms. *Nature*, 446(7133):305–307.

Siebert, C., Rödiger, T., Mallast, U., Gräbe, A., Guttman, J., Laronne, J. B., Storz-Peretz, Y., Greenman, A., Salameh, E., Al-Raggad, M., et al. (2014). Challenges to estimate surface-and groundwater flow in arid regions: The Dead Sea catchment. *Sci. Total Environ.*, 485:828–841.

Smith, R. B. (1985). On severe downslope winds. *Journal of the atmospheric sciences*, 42(23):2597–2603.

Stein, S. and Wysession, M. (2009). *An introduction to seismology, earthquakes, and earth structure.* Wiley-Blackwell, 9th edition. x, 498 pp.

Stull, R. B. (1988). *An introduction to boundary layer meteorology*, volume 1. Atmospheric Sciences Library, Dordrecht: Kluwer, 1988. xii, 666 pp.

Tanimoto, T. and Artru-Lambin, J. (2007). Interaction of solid earth, atmosphere, and ionosphere. *Treatise on Geophysics*, 4:421–444.

Tennekes, H. (1973). The logarithmic wind profile. *J. Atmos. Sci.*, 30(2):234–238. doi: 10.1175/1520-0469(1973)030<0234:TLWP>2.0.CO;2.

Thun, J. (2011). Untersuchung des Einflusses von Wind auf breitbandige seismologische Messungen. Diploma thesis, Karlsruhe Institute of Technology.

Trnkoczy, A., Bormann, P., Hanka, W., Holcomb, L. G., Nigbor, R. L., Shinohara, M., Shiobara, H., and Suyehiro, K. (2012). Site selection, preparation and installation of seismic stations. in: Bormann, P. (Ed.), New Manual of Seismological Observatory Practice 2 (NMSOP-2). *GFZ German Research Centre for Geosciences, Potsdam; http://nmsop.gfz-potsdam.de*, pages 1–139. doi: 10.2312/GFZ.NMSOP-2_ch7.

Tsvieli, Y. and Zangvil, A. (2007). Synoptic climatological analysis of Red Sea Trough and non-Red Sea Trough rain situations over Israel. *Adv. Geosci.*, 12:137–143.

Van der Hoven, I. (1957). Power spectrum of horizontal wind speed in the frequency range from 0.0007 to 900 cycles per hour. *Journal of Meteorology*, 14(2):160–164.

Wapenaar, K. and Fokkema, J. (2006). Green's function representations for seismic interferometry. *Geophysics*, 71(4):SI33–SI46.

Whiteman, C. D. (1990). Observations of thermally developed wind systems in mountainous terrain. *Meteor. Monogr.*, 23(45):5–42.

Widmer, F. (1995). On noise reduction in vertical seismic records below 2 mHz using local barometric pressure w. zürn. *Geophys. Res. Lett.*, 22(24):3537–3540.

Wielandt, E. (2012). Seismic sensors and their calibration. in: Bormann, P. (Ed.), New Manual of Seismological Observatory Practice 2 (NMSOP-2). *GFZ German Research Centre for Geosciences, Potsdam; http://nmsop.gfz-potsdam.de*, pages 1–51. doi: 10.2312/GFZ.NMSOP-2_CH5.

Wilcock, W. S., Webb, S. C., and Bjarnason, I. T. (1999). The effect of local wind on seismic noise near 1 Hz at the MELT site and in Iceland. *Bull. Seismol. Soc. Am.*, 89(6):1543–1557.

Wilczak, J. M., Oncley, S. P., and Stage, S. A. (2001). Sonic anemometer tilt correction algorithms. *Bound.-Lay. Meteorol.*, 99(1):127–150.

Withers, M. M., Aster, R. C., Young, C. J., and Chael, E. P. (1996). High-frequency analysis of seismic background noise as a function of wind speed and shallow depth. *Bull. Seismol. Soc. Am.*, 86(5):1507–1515.

Yechieli, Y., Gavrieli, I., Berkowitz, B., and Ronen, D. (1998). Will the Dead Sea die? *Geology*, 26(8):755–758.

Zhang, J., Gerstoft, P., and Shearer, P. M. (2009). High-frequency P-wave seismic noise driven by ocean winds. *Geophys. Res. Lett.*, 36(9):1–5. L09302.

Zürn, W., Exss, J., Steffen, H., Kroner, C., Jahr, T., and Westerhaus, M. (2007). On reduction of long-period horizontal seismic noise using local barometric pressure. *Geophys. J. Int.*, 171(2):780–796.

List of Figures

List of Tables

Glossary

F_f	Friction force	24
F_p	Dynamic pressure force	24
ν	Dynamic viscosity	22
$\overline{p}$	Mean pressure	22
Φ	Geopotential	22
ρ	Density	24
σ_u^2	Variance of the horizontal wind speed	168
σ_w^2	Variance of the vertical wind speed	168
τ	Shear stress	23
u_i	Wind speed component	21
$\overline{u}_i$	Mean wind speed component	21, 24
u_i'	Turbulent perturbation from mean wind speed	21, 23–25
u_*	Friction velocity	24, 25
$w'T'$	Kinematic temperature flux in the vertical	168
AD	Analogue to digital conversion factor for seismometers	77
$b(t)$	Array beam	29
d_i	Inter station distance perpendicular to the wavefront	27

f_c	Corner frequency	76
$F(s)$	Fourier transform of $f(t)$	224
$f(t)$	Input signal into seismometer	16, 77, 223–225
$G(s)$	Fourier transform of $g(t)$	224
$g(t)$	Output signal of seismometer	16, 77, 223–225
$H(s)$	Transfer function of LTI system, instrument response of a seismometer	224
k	Wave number vector	29
m_c	Factor to convert seismic data from units of counts to nms^{-1}	77
n_i	Noise signal at ith seismological station	28
ω	Complex angular frequency	29
ϕ	Backazimuth	27
r_i	ith location of seismological station	27, 28
S	Ground motion susceptibility: susceptibility of ground motion velocity to wind	89, 159, 162, 178, 185, 186
γ	Seismometer gain	77
s	Seismometer sensitivity	77
θ	Azimuth	27
t_i	Time delay at ith seismological station	27, 28
u	Horizontal slowness	27, 28
v	ground motion velocity	77
v_{app}	Apparent wave velocity	28

ATF	Array transfer function	30, 60, 62, 67
FFT	Fast fourier transform	75, 79–83
iFFT	Inverse fast fourier transform	79
fk-analysis	Frequency-wavenumber analysis	28, 30, 60, 61, 63–67, 141, 142
HPF	High-pass filter	76
LTI	Linear time invariant (system)	16, 223, 224
MS	Collocated, meteorological-seismological dataset	85–88, 90, 186, 192
MSEED	Seismological data format standard miniSEED	75
NHNM	New high noise model, according to Peterson (1993)	58, 182, 183, 187
NLNM	New low noise model, according to Peterson (1993)	10, 58, 182, 183, 187, 188
PBL	Planetary boundary layer	18, 20, 21, 69, 70, 171
PSD	Power spectral density	58, 74, 80–82, 85–91, 94, 101–105, 107, 114–116, 123, 125–132, 134, 135, 148, 152–156, 158, 160–163, 169–177, 180–183, 186–188, 190, 191, 193, 211–213

Acronyms

CTBTO	Comprehensive Nuclear-Test-Ban Treaty Organization	63, 66
DESERVE	Helmholtz Virtual Institute Dead Sea Research Venue	35
GEOFON	GEOFOrschungsNetz	35
GFS	Global Forecasting System	35, 96, 98–100, 107, 108, 110, 111, 114, 118, 119, 132, 138
GFZ	German Research Centre for Geosciences Potsdam	35, 49, 53, 54, 71
GIPP	Geophysical Instrument Pool	49, 54
GPI	Geophysical Institute	35, 40
IMK	Institute of Meteorology and Climate Research	35, 37
KaSP	The Karlsruhe Seismology Processing Toolbox for MATLAB	73, 82
KIT	Karlsruhe Institute of Technology	35, 40
MEMR	Ministry of Energy and Mineral Resources	35, 41, 48
NRA	National Resources Authority in Jordan	41
UFZ	Helmholtz Centre for Environmental Research	35

A Appendix

A.1 Mathematical Description of Seismometer Recordings

The following derivations are to deliver insight into the process of signal acquisition since relations explained here are needed to understand data processing, which is documented in Chapter 5. However, this is rather a short outline, derived from Scherbaum and Schmidtke (2007) and Wielandt (2012) who deduce and describe signal processing routines in seismology in great detail.

Mathematically, a seismometer can be described as a linear time invariant (LTI) system, implicating that the system does not change the waveform of sine waves with constant, exponentially growing, or decaying amplitudes. It can be described by a linear differential equation.

An input signal $f(t)$, such as a seismic wave, produces an output signal $g(t)$ at the seismometer. The signals can be described as eigenfunctions of the differential operators:

$$f(t) = c_1 \cdot e^{(\sigma+i\omega)t} + c_1^* \cdot e^{(\sigma-i\omega)t} \tag{A.1}$$

and

$$g(t) = d_1 \cdot e^{(\sigma+i\omega)t} + d_1^* \cdot e^{(\sigma-i\omega)t}, \tag{A.2}$$

where the * marks the complex conjugate. As signals of definite lengths $f(t)$ and $g(t)$ can be decomposed using the Laplace transform, here illustrated for $f(t)$:

$$f(t) = \frac{1}{2\pi i} \int_{\sigma-i\infty}^{\sigma+i\infty} F(s) e^{st} ds, \tag{A.3}$$

where

$$F(s) = \int_0^\infty f(t)e^{-st}\,dt \tag{A.4}$$

is its Laplace transform $F(s)$.

We can describe the analogue data acquisition of the LTI system, respectively of the seismometer, using a linear differential equation:

$$\sum_{m=0}^{k} c_m f^{(m)}(t) = \sum_{n=0}^{l} d_n g^{(n)}(t), \tag{A.5}$$

where $m,n \in \mathrm{N}$. $f^{(m)}(t)$ and $g^{(n)}(t)$ describe input and output signals and its derivatives assessed by the constants c_m and d_n. When describing the terms of Equation A.5 in the s-domain we get:

$$\sum_{m=0}^{k} c_m s^m F(s) = \sum_{n=0}^{l} d_n s^n G(s), \tag{A.6}$$

where $G(s)$ is the Laplace transform of the output signal $g(t)$. Now, with the system equation in the s-domain, we benefit from the omission of the derivatives of $f(t)$ and $g(t)$. The equation can simply be resolved for the output signal:

$$G(s) = \frac{\sum_{m=0}^{k} c_m s^m}{\sum_{n=0}^{l} d_2 n^n} F(s). \tag{A.7}$$

The term

$$H(s) = \frac{\sum_{m=0}^{k} c_m s^m}{\sum_{n=0}^{l} d_2 n^n} \tag{A.8}$$

is a rational function linking the signal output with the signal input in the s-domain:

$$G(s) = H(s) \cdot F(s). \tag{A.9}$$

The function $H(s)$ is the transfer function of the LTI system, or the instrument response of a seismometer. The instrument acts as a filter to the original input signal $f(t)$. Knowing its response, the primary signal can be calculated from the signal output $g(t)$, as described in Chapter 5.1.2.

A.2 Additional Tables

Table A.1: List of MATLAB codes from foreign authors used for this work. Codes marked with * were modified for present use.

Code	Purpose	Author	Source
`dem.m`*	Shaded relief topography in maps	Francois Beauducel (`beauducel@ipgp.fr`)	http://www.mathworks.com/ matlabcentral/fileexchange/
`rdmseed.m`	Read miniSEED into MATLAB	Francois Beauducel (`beauducel@ipgp.fr`)	http://www.mathworks.com/ matlabcentral/fileexchange/
`longtermspectrogram.m`*	Calculation of long spectrograms	Jörn Groos	KaSP Toolbox GPI
`slidingtimewindowfk.m`	Sliding time window fk-analysis	Tobias Baumann	KaSP Toolbox GPI
`getfkfk.m`*	fk-analysis	Tobias Baumann	KaSP Toolbox GPI
`demeandata.m`	seismic data processing	Jörn Groos	KaSP Toolbox GPI
`detrenddata.m`	seismic data processing	Jörn Groos	KaSP Toolbox GPI
`tapertukeydata.m`	seismic data processing	Jörn Groos	KaSP Toolbox GPI
`filterbuttertimedomain`	seismic data processing	Jörn Groos	KaSP Toolbox GPI
`zerophasedata.m`			KaSP Toolbox GPI
`st.m`*	stockwell transformation short spectrograms	Aditya Sundar	http://www.mathworks.com/ matlabcentral/fileexchange/

Table A.2: List of codes by foreign authors used for this work. Codes marked with * were highly modified for present use.

Code	Purpose	Author	Source
`cube_mseed` (Release 2015,07,24)	Conversion of data from CUBE to MSEED	Trond Ryberg (`trond@gfzÂpotsdam.de`)	http://www.gfz-potsdam.de/en/section/ geophysical-deep-sounding/servicesinfrastructure/ geophysical-instrument-pool-potsdam-gipp/ software/dss-cube/cube2mseed/

Table A.3: Parameters for data streams HHZ, HHN, and HHE from GEOFON Station GHAJ. Source: http://geofon.gfz-potsdam.de/waveform/archive/station.php

Identifier	Sensor	Datalogger	Total Gain	Seismometer S/N
GHAJ	Trillium-240/g=1200 GFZ:GE1993	Q330/HR/g=1677720 GFZ:GE1993	2013264000	639

A.3 Additional Figures

Additional Figures Chapter 4

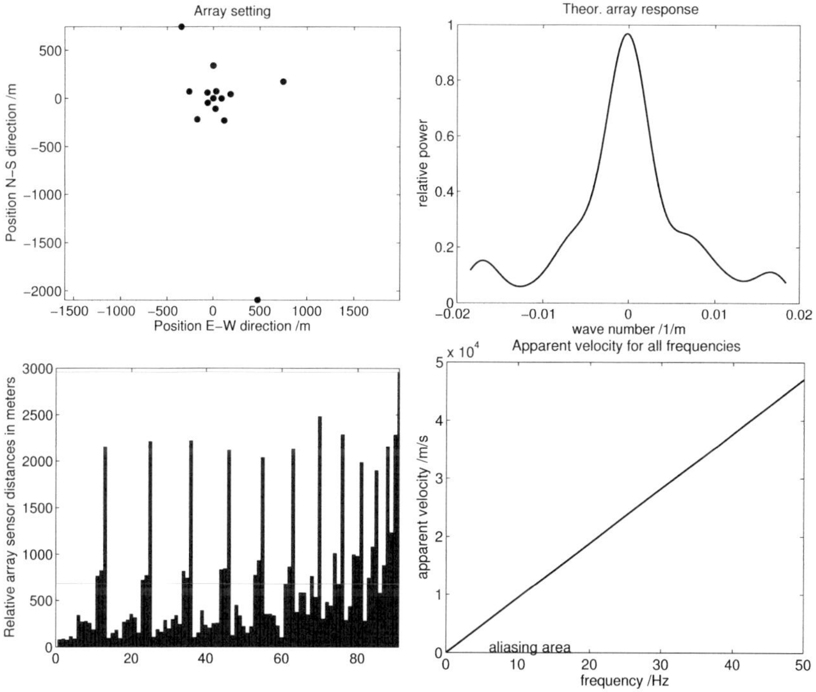

Figure A.1: Array response characteristics. **Top left:** Station Locations. **Top right:** Theoretical array response. **Bottom left:** Relative station distances. **Bottom right:** Apparent velocity with respect to frequency.

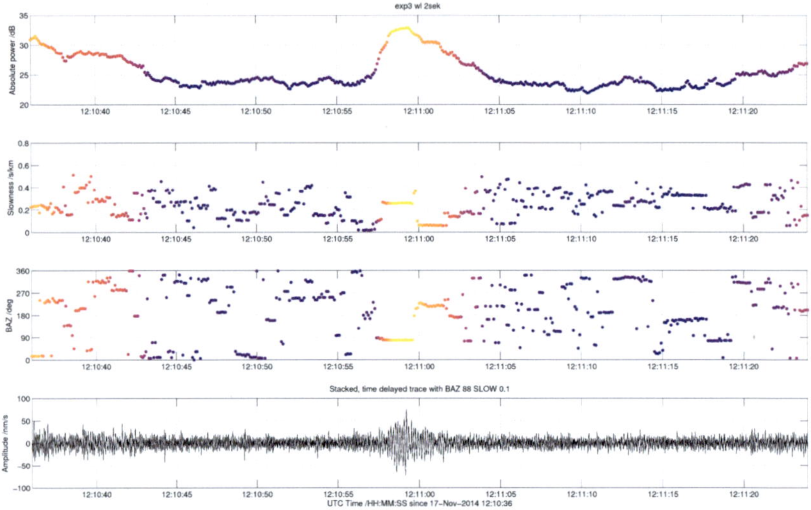

Figure A.2: Sliding time window fk-analysis of CTBTO explosion. Data base: all DES stations except DES14. Length of sliding time window: 1 s, Time steps: 0.05 s, frequency domain: 2 Hz - 18 Hz. All panels are displayed with respect to the same time axis. **a:** Absolute power. **b:** Slowness. **c:** Backazimuth **d:** Seismic velocity, stacked and time-shifted (delay and sum).

Additional Results Chapters 6 and 7

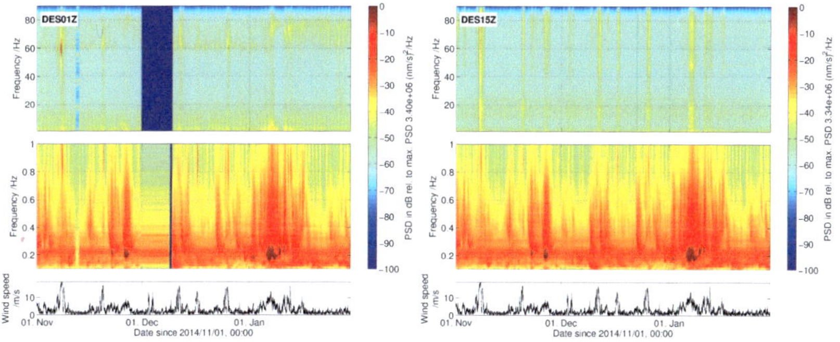

Figure A.3: Spectrograms of seismic velocity from stations DES01 and DES15, Z-component from 01.11.2014 to 31.01.2015

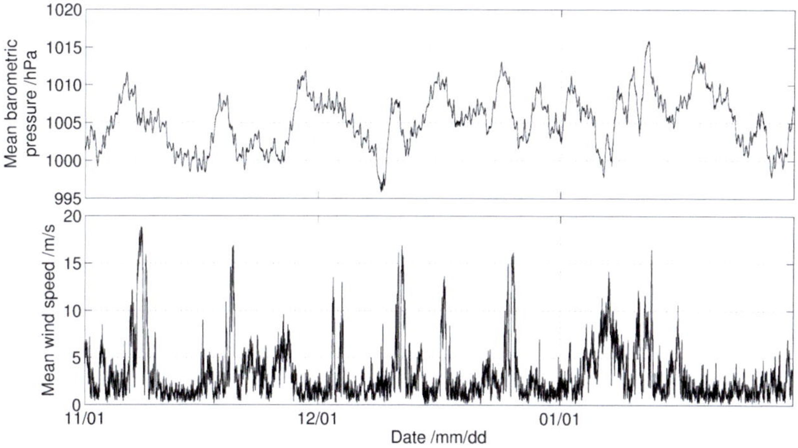

Figure A.4: Mean barometric pressure and mean wind speed at MET1 from 01.11.2014 to 31.01.2015

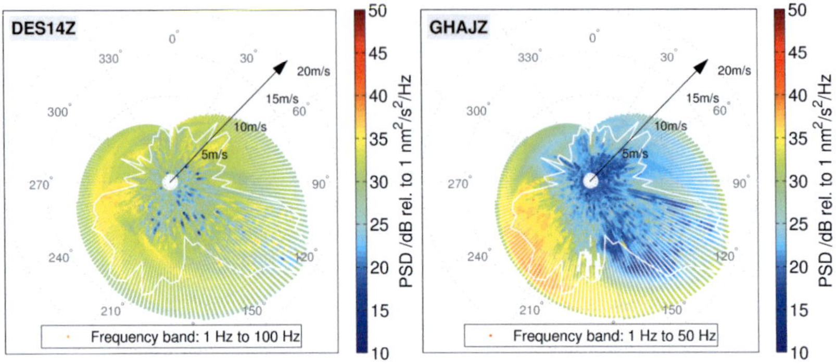

Figure A.5: Polar interpolation of PSD of seismic velocity with respect to wind speed and wind direction at Stations DES14 and GHAJ, Z-components, from 01.11.2014 to 31.01.2015. The PSD is displayed as coloured dots with respect to the wind speed (radial axis) and the azimuth of the wind direction (polar axis). The white, solid line marks the area data is available. Outside the line, points are interpolated without reference to existing data points. Frequency band of PSD data: 1 Hz and 100 Hz.

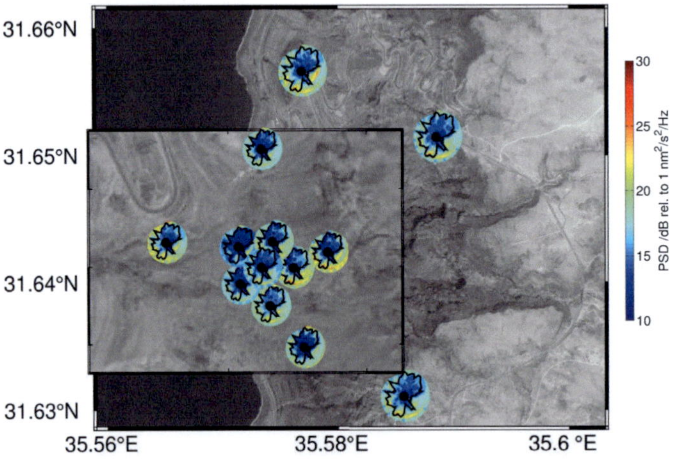

Figure A.6: Polar interpolation of PSD at DES Stations DES01 to DES13 and DES15, with respect to wind speed and wind direction. Polar plots are located in the map at corresponding station locations (cf. Figure 4.2). Colours indicate the PSD with respect to the colour axis valid for all polar plots. Black lines in polar plots mark the data coverage, black dots mark the centres of the polar projections. Period: 01.07.2014 to 31.07.2014.

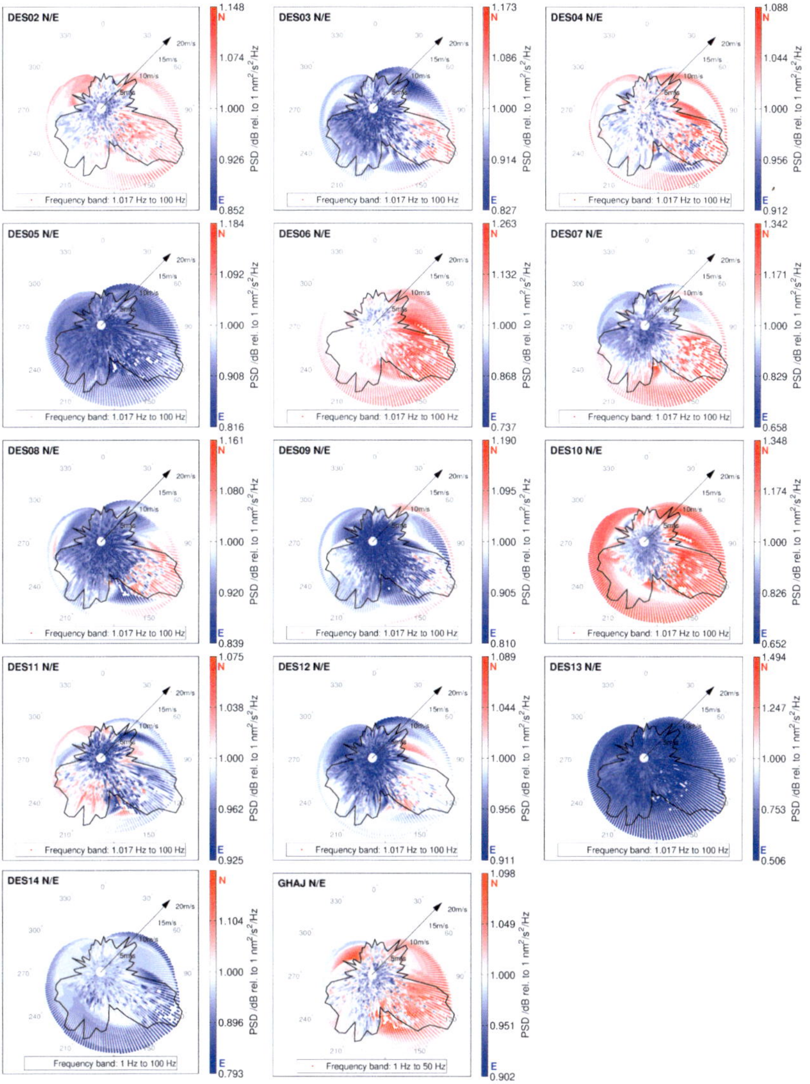

Figure A.7: Polar interpolation of ground velocity PSD of N-component divided by PSD of E-component with respect to wind speed and wind direction at all seismological stations from 01.11.2014 to 31.01.2015. The relation of N-component PSD to E-component PSD is displayed as coloured dots with respect to the wind speed (radial axis) and the azimuth of the wind direction (polar axis). The black, solid line marks the area data is available. Outside the line, points are interpolated without reference to existing data points. Frequency band of PSD data: 1 Hz to 100 Hz.

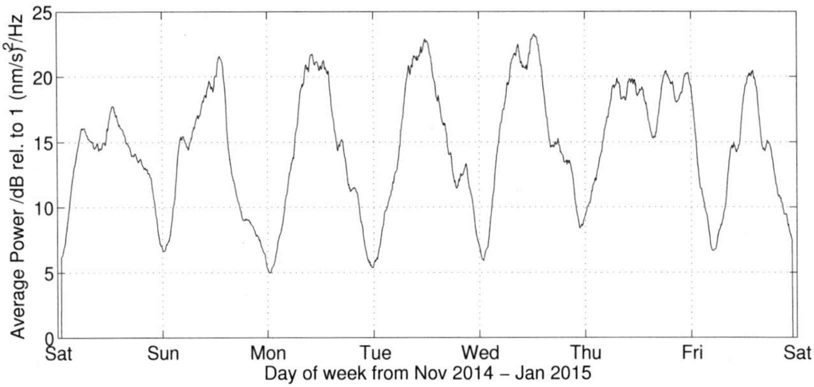

Figure A.8: Weekly variation of mean power time series at all DES Stations except DES14, Z-component. Time series are averaged for all stations, then averaged for weekly cycle. Period: 01.11.2014 to 31.01.2015. Frequency band: above 1 Hz.

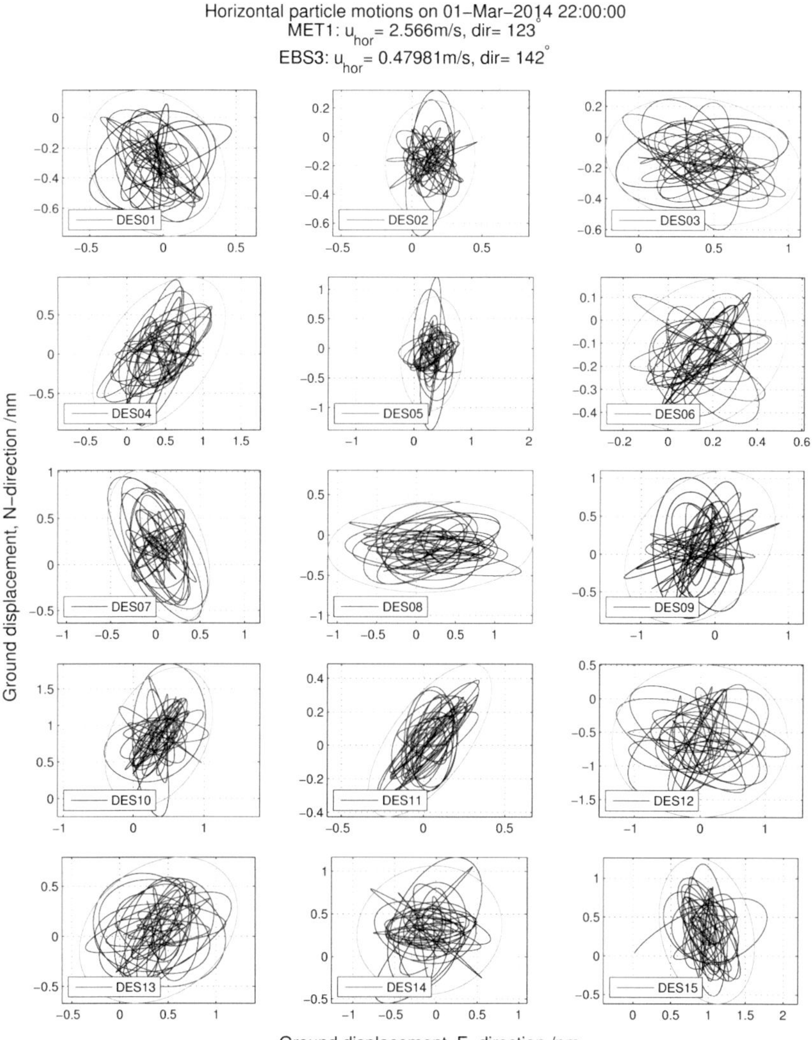

Figure A.9: Horizontal particle motions, ground displacement in N- and E-direction. Period from 01.03.2014, 22:00:00 UTC to 01.03.2014, 22:00:10 UTC. Frequency domain: 2 Hz - 3 Hz.

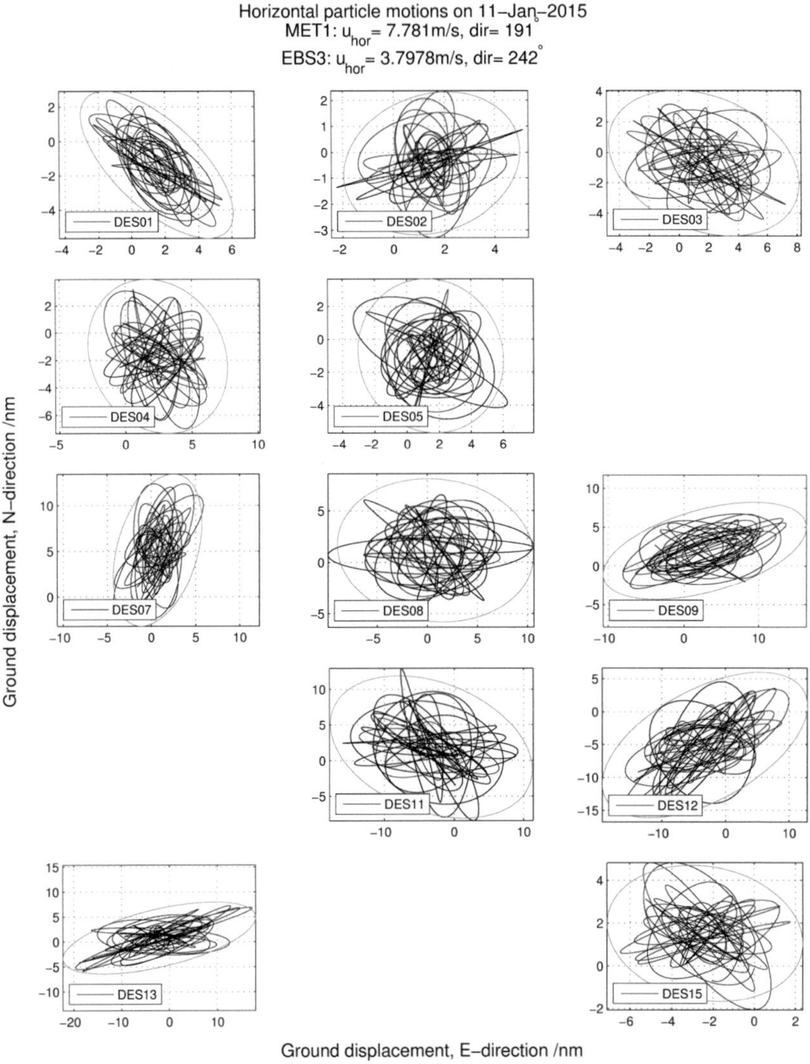

Figure A.10: Horizontal particle motions, ground displacement in N- and E-direction. Period from 11.01.2015, 00:00:00 UTC to 11.01.2015, 00:00:10 UTC. Frequency domain: 2 Hz - 3 Hz.

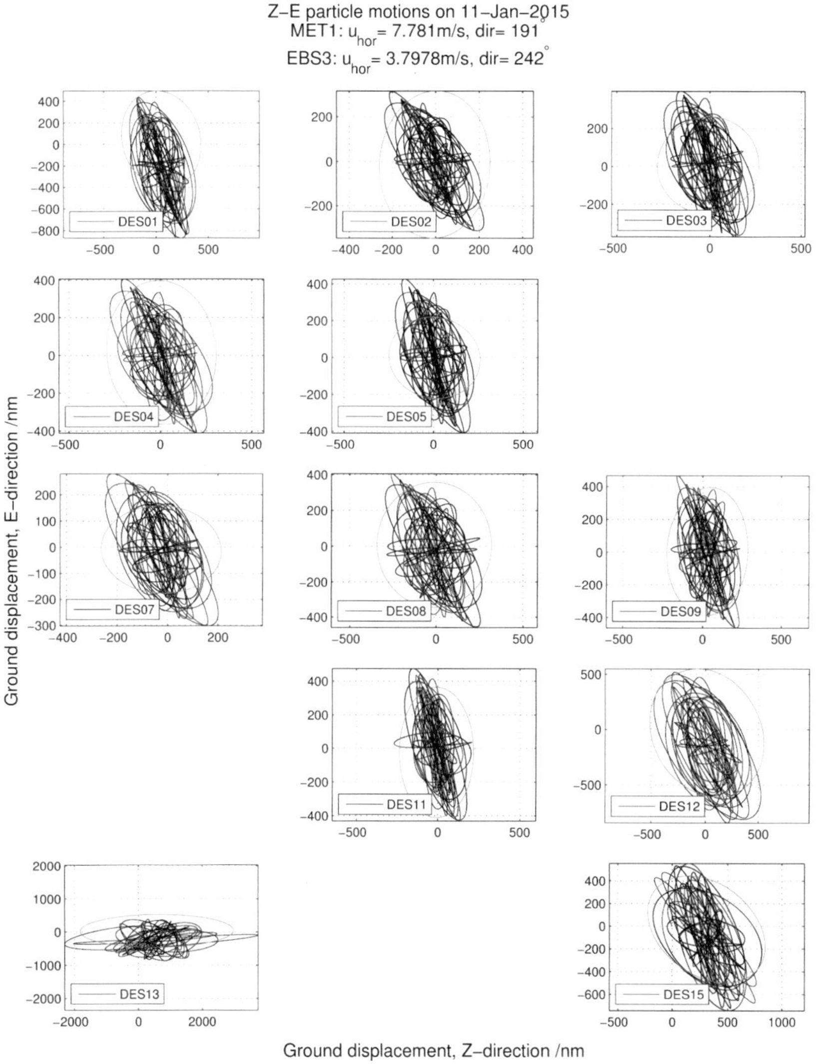

Figure A.11: Vertical particle motions, ground displacement in N- and E-direction. Period from 11.01.2015, 00:00:00 UTC to 11.01.2015, 00:02:00 UTC. Frequency domain: 0.015 Hz - 0.3 Hz.

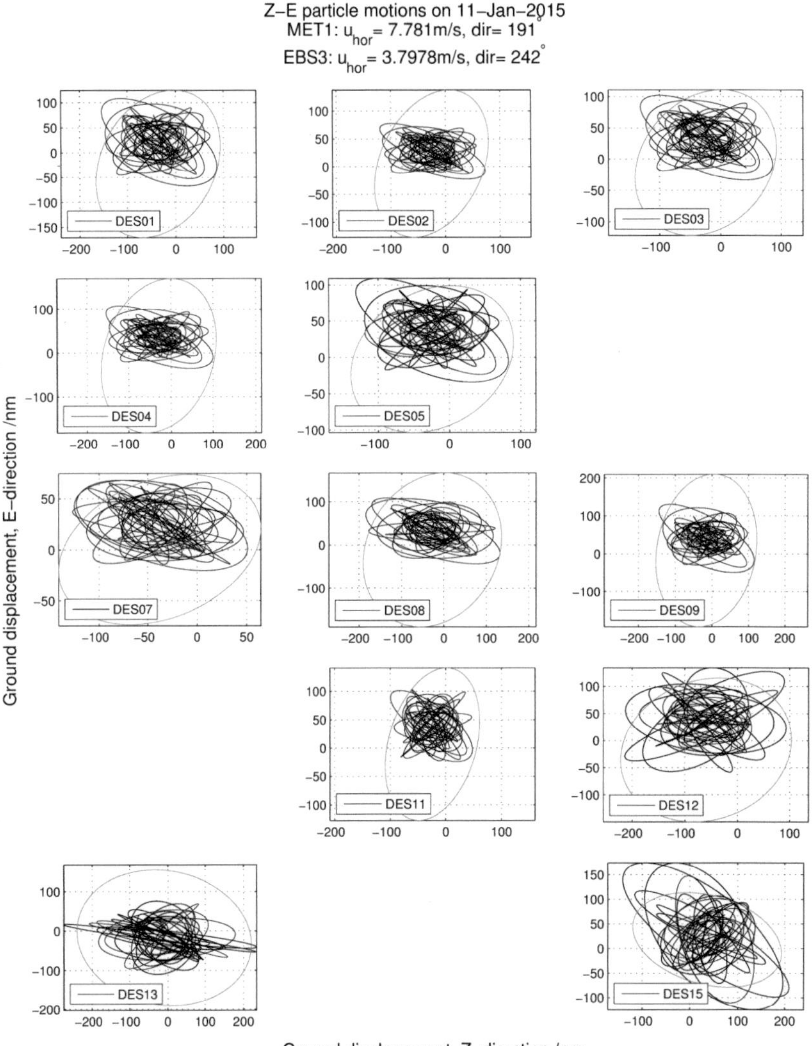

Figure A.12: Vertical particle motions, ground displacement in N- and E-direction. Period from 11.01.2015, 00:00:00 UTC to 11.01.2015, 00:01:00 UTC. Frequency domain: 0.4 Hz - 0.6 Hz.

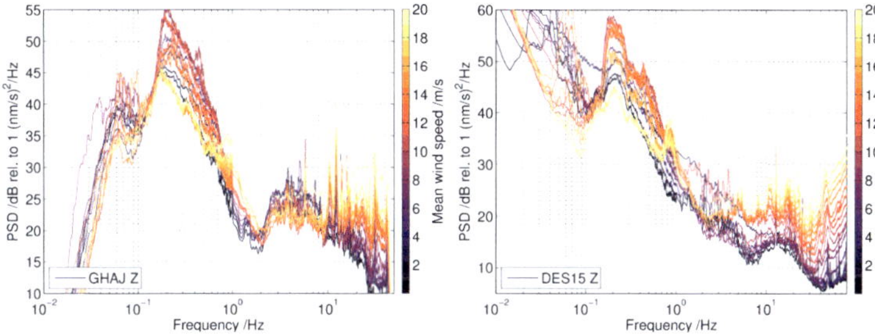

Figure A.13: Noise level of stations GHAJ (left) and DES15 (right), Z-component, from 01.11.2014 to 31.01.2015 depending on the mean wind speed.

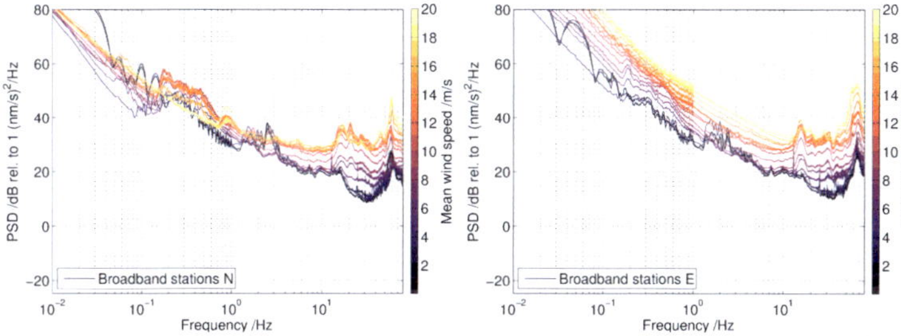

Figure A.14: Noise level of stations DES broad band stations from 01.11.2014 to 31.01.2015 depending on the mean wind speed. **Left:** N-component. **Right:** E-component.

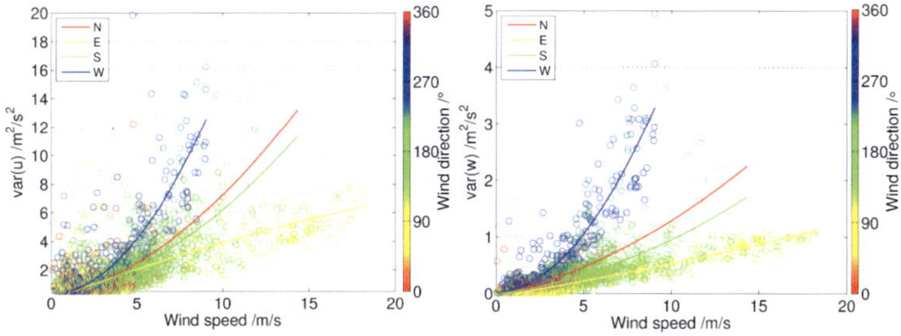

Figure A.15: Variances of the horizontal and vertical turbulent wind from 01.11.2014 to 31.01.2015 at MET1. The colour indicates the mean wind direction. Exponential regression lines are calculated for the four main wind directions. Averaging interval: 30 minutes. **Left:** Variance of the horizontal wind in mean wind direction **Right:** Variances of the vertical wind.

A.4 Instrument Information and Experiment Documentation

```
**********************************************************************
*                                                                    *
*         Configuration File for the 1 & 3 Channel DSS-Cubes         *
*                                                                    *
*                             Software V0.11T                        *
**********************************************************************
* Syntax:                                                            *
*  - Lines beginning with a star '*' are comments (and are ignored) *
*  - Empty lines are not allowed                                     *
*  - Parameter keywords are of six capitalized characters,           *
*    directly followed by an equal sign                              *
*  - The parameter value starts at column #8                         *
*  - Line length is limited to 70 characters                         *
**********************************************************************
****** !!! Parameters for 3 channel Cubes only !!!            ******
**********************************************************************
*
** Active Channels       1 = ch1  // 2 = ch1 & ch2  // 3 = ch1 to ch3
CH_NUM=3
*
** Amplifier Gain        1, 2, 4, 8, 16, 32 or 64
P_AMPL=1
*
** AD Converter Mode     0 = Low power // 1 = High resolution
C_MODE=0
*
** Amplifier Chopping    0 = off // 1 = on
A_CHOP=1
*
**********************************************************************
*****   General parameters for 1 and 3 channel Cubes             *****
**********************************************************************
*
** Project Name          Project name of max. 20 characters
E_NAME=JORverGFZ
*
** Sample Rate           50, 100, 200, 400, (800@1ch) sps
S_RATE=200
*
** Digital High Pass     0 = off // 1 = on
A_FILT=0
*
** FIR Filter            0 = Linear phase // 1 = Minimum phase
A_PHAS=0
*
** TimeBase Correction   0 = off // 1 = PLL // 2 = DIFF
PLL_XO=0
*
** Geographic position lat, lon, alt; to speed up satellite search
GPS_PO=>SIP+31+035+0000
*
** GPS Mode              0 = cycled // 1 = continuously
GPS_ON=0
*
** GPS & Flush Interval 3 to 60 minutes
*  (GPS interval only in cycled mode)
F_TIME=30
*
**********************************************************************
*****   Parameters relevant for GPS cycled mode only             *****
**********************************************************************
** GPS OFF after         0 = 'GPS_TI' // 1 = 60 GPS fixes max.'GPS_TI'
GPS_OF=0
*
** GPS ON Time           3 to 60 minutes
GPS_TI=5
```

Scripture A.16: Config file of CUBE logger 640

Instrument Responses

```
[Vs/m]
MARK L-4C-3D; No. 1354a ; E-W Component
@ Recorder Input of 100000 Ohm
15.12.2009
 1.84825454711914E+0002
2
-4.55033683776855E+0000   4.49520492553711E+0000
-4.55033683776855E+0000  -4.49520492553711E+0000
2
 0.00000000000000E+0000   0.00000000000000E+0000
 0.00000000000000E+0000   0.00000000000000E+0000
1 @ 10 Hz

[Vs/m]
MARK L-4C-3D; No. 1354a ; N-S Component
@ Recorder Input of 100000 Ohm
15.12.2009
 1.78155273437500E+0002
2
-4.48076725006104E+0000   4.00986194610596E+0000
-4.48076725006104E+0000  -4.00986194610596E+0000
2
 0.00000000000000E+0000   0.00000000000000E+0000
 0.00000000000000E+0000   0.00000000000000E+0000
1 @ 10 Hz

[Vs/m]
MARK L-4C-3D; No. 1354a ; Z Component
@ Recorder Input of 100000 Ohm
15.12.2009
 1.77282333374023E+0002
2
-4.05956840515137E+0000   6.50968122482300E+0000
-4.05956840515137E+0000  -6.50968122482300E+0000
2
 0.00000000000000E+0000   0.00000000000000E+0000
 0.00000000000000E+0000   0.00000000000000E+0000
1 @ 10 Hz
```

Scripture A.17: Poles and Zeros M1354a

```
[Vs/m]
MARK L-4C-3D; No. 1355a ; E-W Component
@ Recorder Input of 100000 Ohm
15.12.2009
 1.86128890991211E+0002
2
-4.63865184783936E+0000   4.61139011383057E+0000
-4.63865184783936E+0000  -4.61139011383057E+0000
2
 0.00000000000000E+0000   0.00000000000000E+0000
 0.00000000000000E+0000   0.00000000000000E+0000
1 @ 10 Hz

[Vs/m]
MARK L-4C-3D; No. 1355a ; N-S Component
@ Recorder Input of 100000 Ohm
15.12.2009
 1.76486343383789E+0002
2
-4.55630016326904E+0000   4.24395227432251E+0000
-4.55630016326904E+0000  -4.24395227432251E+0000
2
 0.00000000000000E+0000   0.00000000000000E+0000
 0.00000000000000E+0000   0.00000000000000E+0000
1 @ 10 Hz

[Vs/m]
MARK L-4C-3D; No. 1355a ; Z Component
@ Recorder Input of 100000 Ohm
15.12.2009
 1.76397521972656E+0002
2
-4.38063907623291E+0000   6.16753959655762E+0000
-4.38063907623291E+0000  -6.16753959655762E+0000
2
 0.00000000000000E+0000   0.00000000000000E+0000
 0.00000000000000E+0000   0.00000000000000E+0000
1 @ 10 Hz
```

Scripture A.18: Poles and Zeros M1355a

```
[Vs/m]
MARK L-4C-3D; No. 1357a ; E-W Component
@ Recorder Input of 100000 Ohm
15.12.2009
 1.84042816162109E+0002
2
-4.47165584564209E+0000   4.38704872131348E+0000
-4.47165584564209E+0000  -4.38704872131348E+0000
2
 0.00000000000000E+0000   0.00000000000000E+0000
 0.00000000000000E+0000   0.00000000000000E+0000
1 @ 10 Hz

[Vs/m]
MARK L-4C-3D; No. 1357a ; N-S Component
@ Recorder Input of 100000 Ohm
15.12.2009
 1.73393875122070E+0002
2
-4.60344457626343E+0000   4.58441781997681E+0000
-4.60344457626343E+0000  -4.58441781997681E+0000
2
 0.00000000000000E+0000   0.00000000000000E+0000
 0.00000000000000E+0000   0.00000000000000E+0000
1 @ 10 Hz

[Vs/m]
MARK L-4C-3D; No. 1357a ; Z Component
@ Recorder Input of 100000 Ohm
15.12.2009
 1.63335159301758E+0002
2
-4.22283172607422E+0000   7.50098228454590E+0000
-4.22283172607422E+0000  -7.50098228454590E+0000
2
 0.00000000000000E+0000   0.00000000000000E+0000
 0.00000000000000E+0000   0.00000000000000E+0000
1 @ 10 Hz
```

Scripture A.19: Poles and Zeros M1357a

```
[Vs/m]
MARK L-4C-3D; No. 1823_ ; E-W Component
@ Recorder Input of 100000 Ohm
15.12.2009
 1.64581192016602E+0002
2
-4.53893423080444E+0000   5.13697242736816E+0000
-4.53893423080444E+0000  -5.13697242736816E+0000
2
 0.00000000000000E+0000   0.00000000000000E+0000
 0.00000000000000E+0000   0.00000000000000E+0000
1 @ 10 Hz

[Vs/m]
MARK L-4C-3D; No. 1823_ ; N-S Component
@ Recorder Input of 100000 Ohm
15.12.2009
 1.74039245605469E+0002
2
-4.50291919708252E+0000   4.60442495346069E+0000
-4.50291919708252E+0000  -4.60442495346069E+0000
2
 0.00000000000000E+0000   0.00000000000000E+0000
 0.00000000000000E+0000   0.00000000000000E+0000
1 @ 10 Hz

[Vs/m]
MARK L-4C-3D; No. 1823_ ; Z Component
@ Recorder Input of 100000 Ohm
15.12.2009
 1.69862655639648E+0002
2
-4.36489486694336E+0000   6.18638372421265E+0000
-4.36489486694336E+0000  -6.18638372421265E+0000
2
 0.00000000000000E+0000   0.00000000000000E+0000
 0.00000000000000E+0000   0.00000000000000E+0000
1 @ 10 Hz
```

Scripture A.20: Poles and Zeros M1823

```
[Vs/m]
MARK L-4C-3D; No. 1824_ ; E-W Component
@ Recorder Input of 100000 Ohm
15.12.2009
 1.80617263793945E+0002
2
-4.55526590347290E+0000   4.19884634017944E+0000
-4.55526590347290E+0000  -4.19884634017944E+0000
2
 0.00000000000000E+0000   0.00000000000000E+0000
 0.00000000000000E+0000   0.00000000000000E+0000
1 @ 10 Hz

[Vs/m]
MARK L-4C-3D; No. 1824_ ; N-S Component
@ Recorder Input of 100000 Ohm
15.12.2009
 1.62330398559570E+0002
2
-4.53122282028198E+0000   4.96657419204712E+0000
-4.53122282028198E+0000  -4.96657419204712E+0000
2
 0.00000000000000E+0000   0.00000000000000E+0000
 0.00000000000000E+0000   0.00000000000000E+0000
1 @ 10 Hz

[Vs/m]
MARK L-4C-3D; No. 1824_ ; Z Component
@ Recorder Input of 100000 Ohm
15.12.2009
 1.71045745849609E+0002
2
-5.51097106933594E+0000   6.00334453582764E+0000
-5.51097106933594E+0000  -6.00334453582764E+0000
2
 0.00000000000000E+0000   0.00000000000000E+0000
 0.00000000000000E+0000   0.00000000000000E+0000
1 @ 10 Hz
```

Scripture A.21: Poles and Zeros M1824

```
[Vs/m]
MARK L-4C-3D; No. 1871_ ; E-W Component
@ Recorder Input of 100000 Ohm
15.12.2009
 1.71083007812500E+0002
2
-4.52703094482422E+0000   4.69486665725708E+0000
-4.52703094482422E+0000  -4.69486665725708E+0000
2
 0.00000000000000E+0000   0.00000000000000E+0000
 0.00000000000000E+0000   0.00000000000000E+0000
1 @ 10 Hz

[Vs/m]
MARK L-4C-3D; No. 1871_ ; N-S Component
@ Recorder Input of 100000 Ohm
15.12.2009
 1.72439971923828E+0002
2
-4.66227483749390E+0000   4.32374572753906E+0000
-4.66227483749390E+0000  -4.32374572753906E+0000
2
 0.00000000000000E+0000   0.00000000000000E+0000
 0.00000000000000E+0000   0.00000000000000E+0000
1 @ 10 Hz

[Vs/m]
MARK L-4C-3D; No. 1871_ ; Z Component
@ Recorder Input of 100000 Ohm
15.12.2009
 1.79329925537109E+0002
2
-4.38748216629028E+0000   6.40027141571045E+0000
-4.38748216629028E+0000  -6.40027141571045E+0000
2
 0.00000000000000E+0000   0.00000000000000E+0000
 0.00000000000000E+0000   0.00000000000000E+0000
1 @ 10 Hz
```

Scripture A.22: Poles and Zeros M1871

```
[Vs/m]
MARK L-4C-3D; No. 1879_ ; E-W Component
@ Recorder Input of 100000 Ohm
15.12.2009
 1.77014678955078E+0002
2
-4.56755638122559E+0000   4.36932563781738E+0000
-4.56755638122559E+0000  -4.36932563781738E+0000
2
 0.00000000000000E+0000   0.00000000000000E+0000
 0.00000000000000E+0000   0.00000000000000E+0000
1 @ 10 Hz

[Vs/m]
MARK L-4C-3D; No. 1879_ ; N-S Component
@ Recorder Input of 100000 Ohm
15.12.2009
 1.78521392822266E+0002
2
-4.56095075607300E+0000   4.60903358459473E+0000
-4.56095075607300E+0000  -4.60903358459473E+0000
2
 0.00000000000000E+0000   0.00000000000000E+0000
 0.00000000000000E+0000   0.00000000000000E+0000
1 @ 10 Hz

[Vs/m]
MARK L-4C-3D; No. 1879_ ; Z Component
@ Recorder Input of 100000 Ohm
15.12.2009
 1.65948425292969E+0002
2
-4.34231615066528E+0000   7.83726406097412E+0000
-4.34231615066528E+0000  -7.83726406097412E+0000
2
 0.00000000000000E+0000   0.00000000000000E+0000
 0.00000000000000E+0000   0.00000000000000E+0000
1 @ 10 Hz
```

Scripture A.23: Poles and Zeros M1879

```
[Vs/m]
MARK L-4C-3D; No. 1885_ ; E-W Component
@ Recorder Input of 100000 Ohm
15.12.2009
 1.62591094970703E+0002
2
-4.53695726394653E+0000   4.98684930801392E+0000
-4.53695726394653E+0000  -4.98684930801392E+0000
2
 0.00000000000000E+0000   0.00000000000000E+0000
 0.00000000000000E+0000   0.00000000000000E+0000
1 @ 10 Hz

[Vs/m]
MARK L-4C-3D; No. 1885_ ; N-S Component
@ Recorder Input of 100000 Ohm
15.12.2009
 1.57281753540039E+0002
2
-4.26398897171021E+0000   6.00377559661865E+0000
-4.26398897171021E+0000  -6.00377559661865E+0000
2
 0.00000000000000E+0000   0.00000000000000E+0000
 0.00000000000000E+0000   0.00000000000000E+0000
1 @ 10 Hz

[Vs/m]
MARK L-4C-3D; No. 1885_ ; Z Component
@ Recorder Input of 100000 Ohm
15.12.2009
 1.56050384521484E+0002
2
-4.33532953262329E+0000   5.99904012680054E+0000
-4.33532953262329E+0000  -5.99904012680054E+0000
2
 0.00000000000000E+0000   0.00000000000000E+0000
 0.00000000000000E+0000   0.00000000000000E+0000
1 @ 10 Hz
```

Scripture A.24: Poles and Zeros M1885

```
[Vs/m]
MARK L-4C-3D; No. 1898_ ; E-W Component
@ Recorder Input of 100000 Ohm
15.12.2009
 1.68138198852539E+0002
2
-4.59143114089966E+0000   4.38981580734253E+0000
-4.59143114089966E+0000  -4.38981580734253E+0000
2
 0.00000000000000E+0000   0.00000000000000E+0000
 0.00000000000000E+0000   0.00000000000000E+0000
1 @ 10 Hz

[Vs/m]
MARK L-4C-3D; No. 1898_ ; N-S Component
@ Recorder Input of 100000 Ohm
15.12.2009
 1.68445983886719E+0002
2
-4.56770658493042E+0000   4.35096931457520E+0000
-4.56770658493042E+0000  -4.35096931457520E+0000
2
 0.00000000000000E+0000   0.00000000000000E+0000
 0.00000000000000E+0000   0.00000000000000E+0000
1 @ 10 Hz

[Vs/m]
MARK L-4C-3D; No. 1898_ ; Z Component
@ Recorder Input of 100000 Ohm
15.12.2009
 1.65056243896484E+0002
2
-4.31715536117554E+0000   6.16617822647095E+0000
-4.31715536117554E+0000  -6.16617822647095E+0000
2
 0.00000000000000E+0000   0.00000000000000E+0000
 0.00000000000000E+0000   0.00000000000000E+0000
1 @ 10 Hz
```

Scripture A.25: Poles and Zeros M1898

```
[Vs/m]
MARK L-4C-3D; No. 2831_ ; E-W Component
@ Recorder Input of 100000 Ohm
15.12.2009
 1.68579803466797E+0002
2
-4.47811079025269E+0000   4.26283264160156E+0000
-4.47811079025269E+0000  -4.26283264160156E+0000
2
 0.00000000000000E+0000   0.00000000000000E+0000
 0.00000000000000E+0000   0.00000000000000E+0000
1 @ 10 Hz

[Vs/m]
MARK L-4C-3D; No. 2831_ ; N-S Component
@ Recorder Input of 100000 Ohm
15.12.2009
 1.55674255371094E+0002
2
-4.53529739379883E+0000   4.64306592941284E+0000
-4.53529739379883E+0000  -4.64306592941284E+0000
2
 0.00000000000000E+0000   0.00000000000000E+0000
 0.00000000000000E+0000   0.00000000000000E+0000
1 @ 10 Hz

[Vs/m]
MARK L-4C-3D; No. 2831_ ; Z Component
@ Recorder Input of 100000 Ohm
15.12.2009
 1.66858566284180E+0002
2
-5.09177732467651E+0000   5.37190580368042E+0000
-5.09177732467651E+0000  -5.37190580368042E+0000
2
 0.00000000000000E+0000   0.00000000000000E+0000
 0.00000000000000E+0000   0.00000000000000E+0000
1 @ 10 Hz
```

Scripture A.26: Poles and Zeros M2831

```
Gs = 750 V/(m/s)

3 Zeros
z1 = 0
z2 = 0
z3 = -434.1

7 Poles
p1 = -0.03691   0.03712i
p2 = -0.03691  -0.03712i
p3 = -371.2
p4 = -373.9   475.5i
p5 = -373.9  -475.5i
p6 = -588.4   1508i
p7 = -588.4  -1508i

Normalisation Factor (@1Hz)
a = 8.184*10^11
```

Scripture A.27: Poles and Zeros of Trillium Compact

Working protocol

Table A.4: Working protocol of seismological stations. Sensor drift could not be observed for stations DES06, 10, and 14 due to untimely dismount of the stations by local police.

Station	Sensor, ID	Logger, ID	Sensor drift
DES01	Trillium Compact 015	CUBE 710	no drift
DES02	Mark Products 1357	CUBE 645	drift
DES03	Mark Products 1824	CUBE 641	drift
DES04	Mark Products 1885	CUBE 642	slight drift
DES05	Mark Products 1871	CUBE 643	drift
DES06	Mark Products 1354A	CUBE 644	not observed
DES07	Mark Products 1879	CUBE 640	drift
DES08	Mark Products 1823	CUBE 646	slight drift
DES09	Mark Products 1355A	CUBE 647	drift
DES10	Mark Products 2831	CUBE 648	not observed
DES11	Mark Products 1898	CUBE 649	no drift
DES12	Trillium Compact 013	CUBE 711	slight drift
DES13	Trillium Compact 016	CUBE 712	slight drift
DES14	Trillium Compact 014	CUBE 713	not observed
DES15	Trillium Compact 007	CUBE 650	water bubble broken

Parameters of Meteorological Stations

```
Temperature and Humidity measurement at 2 m height
--------------------------------------------------

Rotronic HygroClip HC2S3, MET21 Radiation Shield
Temperature Range: -40 °C .. +60 °C
Accuracy: ±0.1 °C, long term<0.1°C / year
Humidity Range: 0 .. 100 %
Accuracy: ±0.8%, long term < 1% / year
Data Rate: 0.2 Hz, Averaging Interval: 10 Minutes

3-Component Wind Measurements (u,v,w) at 10 m height
----------------------------------------------------

Gill „WindMaster“ – 1590-PK-020 Ultrasonic Anemometer
Wind Speed Range 0 .. 45 m/s
Wind Speed Accuracy < 1.5% RMS @ 12 m/s
Wind Direction Range 0 .. 359°
Wind Direction Accuracy: 0.5°
Data Rate 20 Hz, Averaging Interval: 10 Minutes

Barometric Pressure Measurement
-------------------------------

Vaisala PTB330 Digital Barometer with Young 61002 Pressure Port
The Pressure Port minimizes dynamic pressure errors due to wind.
Pressure Range: 500 .. 1100 hPa
Accuracy: ±0.25 hPa, long term: ±0.1 hPa / year
Data Rate: 1 Hz, Averaging Interval: 10 Minutes

Precipitation Measurement
-------------------------

Tipping Bucket Raingauge Young 52202
Catchment Area: 200 cm²
Resolution: 0.1 mm
Accuracy: 2% up to to 25 mm/h, 3% up to 50 mm/h
Counting Interval: 10 Minutes

Solar Radiation Measurement
---------------------------

Kipp & Zonen CNR4 Net Radiometer
Pyranometer Spectral Response: 0.35 µm .. 2.8 µm
Pyrgeometer Spectral Response: 4.5 µm .. 42 µm
Pyranometer Accuracy: <5%
Pyrgeometer Accuracy: <10%
Data Rate: 0.2 Hz, Averaging Interval: 10 Minutes

Soil Temperature Measurement
----------------------------

Campbell TCAV Probe
Averages 4 type E thermocouples
Sensor depths 2 cm and 6 cm
Data Rate: 0.2 Hz, Averaging Interval: 10 Minutes

Soil Heat Flux Measurement
--------------------------

Two Hukseflux HFP01 Heat Flux Plates
Range: ±2000 W/m²
Accuracy: -15% .. +5% in most common soils
Data Rate: 0.2 Hz, Averaging Interval: 10 Minutes

Soil Moisture Measurement (volumetric water content)
----------------------------------------------------

Two Campbell CS616 Water Content Reflectometers
Accuracy: ±2.5% VWC in range 0% .. 50% VWC
Data Rate: 0.2 Hz, Averaging Interval: 10 Minutes

Data Acquisition
----------------

Campbell Scientific CR3000 Datalogger with NL115 Ethernet Interface and
Compact Flash Module
Compact Flash card 16 GB

Power Supply
------------

Phaesun Solar Panel, 60 Wpeak, Campbell CH200 Charging Regulator, VRLA
Battery 24 Ah

GPS Time synchronisation (done hourly with 0.2 ms accuracy)
-----------------------------------------------------------

Garmin GPS16X-HVS GPS Receiver (12 channel)
PPS Accuracy: ±1µs
Acquisition Times (Approximate):
Reacquisition: Less than 2 s
Hot: 1 s (all data known)
Warm: ~38 s (initial position, time and almanac known, ephemeris unknown)
Cold: ~45 s

Data Transmission (only Averages)
---------------------------------

Campbell Scientific COM110 GPRS Modem
Quadband, GPRS Multislot Class 10 (< 40 kBit/s Upload)
```

Scripture A.28: Meteorological parameters and instruments at meteorological station MET1

```
Temperature and Humidity measurement at 2 m height
--------------------------------------------------

Rotronic HygroClip HC2S3, MET21 Radiation Shield
Temperature Range: -40 °C .. +60 °C
Accuracy: ±0.1 °C, long term<0.1°C / year
Humidity Range: 0 .. 100 %
Accuracy: ±0.8%, long term < 1% / year
Data Rate: 0.2 Hz, Averaging Interval: 10 Minutes

3-Component Wind Measurements (u,v,w); Water vapour, carbon dioxide at 6 m
height
-------------------------------------

Campbell IRGASON Integrated CO2 /H2O Open-Path Gas Analyzer and 3D Sonic
Anemometer
Wind Speed Range 0 .. 65 m/s
Wind Speed Accuracy (x,y)< ±0.08 m/s
Wind Speed Accuracy (z)< ±0.04 m/s
Wind Direction Range 0 .. 359°
Wind Direction Accuracy:
Wind vector ±5° horizontal: < ±2% of reading
Wind vector ±10° horizontal: < ±3% of reading
Wind vector ±20° horizontal: < ±6% of reading
Data Rate 20 Hz, Averaging Interval: 10 Minutes

Barometric Pressure Measurement
-------------------------------

Vaisala PTB330 Digital Barometer with Young 61002 Pressure Port
The Pressure Port minimizes dynamic pressure errors due to wind.
Pressure Range: 500 .. 1100 hPa
Accuracy: ±0.25 hPa, long term: ±0.1 hPa / year
Data Rate: 1 Hz, Averaging Interval: 10 Minutes

Precipitation Measurement
-------------------------

Tipping Bucket Raingauge Young 52202
Catchment Area: 200 cm²
Resolution: 0.1 mm
Accuracy: 2% up to to 25 mm/h, 3% up to 50 mm/h
Counting Interval: 10 Minutes

Solar Radiation Measurement
---------------------------

Kipp & Zonen CNR4 Net Radiometer
Pyranometer Spectral Response: 0.35 µm .. 2.8 µm
Pyrgeometer Spectral Response: 4.5 µm .. 42 µm
Pyranometer Accuracy: <5%
Pyrgeometer Accuracy: <10%
Data Rate: 0.2 Hz, Averaging Interval: 10 Minutes

Soil Temperature Measurement
----------------------------

Campbell TCAV Probe
Averages 4 type E thermocouples
Sensor depths 2 cm and 6 cm
Data Rate: 0.2 Hz, Averaging Interval: 10 Minutes

Soil Heat Flux Measurement
--------------------------

Two Hukseflux HFP01 Heat Flux Plates
Range: ±2000 W/m²
Accuracy: -15% .. +5% in most common soils
Data Rate: 0.2 Hz, Averaging Interval: 10 Minutes

Soil Moisture Measurement (volumetric water content)
----------------------------------------------------

Two Campbell CS616 Water Content Reflectometers
Accuracy: ±2.5% VWC in range 0% .. 50% VWC
Data Rate: 0.2 Hz, Averaging Interval: 10 Minutes

Data Acquisition
----------------

Campbell Scientific CR3000 Datalogger with NL115 Ethernet Interface and
Compact Flash Module
Compact Flash card 16 GB

Power Supply
------------

Phaesun Solar Panel, 60 Wpeak, Campbell CH200 Charging Regulator, VRLA
Battery 84 Ah

GPS Time synchronisation (done hourly with 0.2 ms accuracy)
-----------------------------------------------------------

Garmin GPS16X-HVS GPS Receiver (12 channel)
PPS Accuracy: ±1µs
Acquisition Times (Approximate):
Reacquisition: Less than 2 s
Hot: 1 s (all data known)
Warm: ~38 s (initial position, time and almanac known, ephemeris unknown)
Cold: ~45 s

Data Transmission (only Averages)
---------------------------------

Campbell Scientific COM110 GPRS Modem
Quadband, GPRS Multislot Class 10 (< 40 kBit/s Upload)
```

Scripture A.29: Meteorological parameters and instruments at meteorological station EBS3

Dank

Die vorliegende Doktorarbeit wurde von 2012 bis 2016 am Institut für Meteorologie und Klimaforschung am Karlsruher Institut für Technologie durchgeführt. Finanziert und gefördert wurde die Arbeit im Rahmen des HGF-Projekts DESERVE. Darüber hinaus wurde sie durch freundliche Kooperation mit dem Geophysikalischen Institut des Karlsruher Institut für Technologie unterstützt. Am Entstehungsprozess dieser Arbeit waren wegen des experimentellen Charakters, bzw. der Planung und Durchführung von Messungen viele Menschen beteiligt, ohne die die Arbeit niemals möglich gewesen wäre. Dem kann man vermutlich nicht gerecht werden, dennoch wage ich einen Versuch.

Großer Dank gilt Herrn Professor Kottmeier, der als Referent und Institutsleiter meine Arbeit maßgeblich betreut und ermöglicht hat. Unter seiner Betreuung durfte ich meine Arbeit selbständig gestalten, wobei mir Herr Kottmeier jederzeit unterstützend zur Seite stand.
Mein besonderer Dank gilt außerdem PD Dr. Joachim Ritter, der nicht nur das Korreferat für meine Arbeit übernahm, sondern weit über das übliche Maß hinaus meine Arbeit bereits von der Entwicklung der Zielsetzung an mitbetreute. Danke für deine beständige, fachliche und persönliche Unterstützung und die Möglichkeit, immer wieder für Diskussionen zu dir und ins Seismo-Seminar zu kommen!
Entscheidend für meine Arbeit war außerdem die Zusammenarbeit mit meinem Betreuer und Arbeitsgruppenleiter Dr. Ulrich Corsmeier. Zusammen mit Herrn Kottmeier und Joachim Ritter ermöglichte er diese Dissertation erst und trotz des - aus meteorologischer Sicht - exotischen Themas, schenkte er mir sowohl fachlich, als auch in Planung und Durchführung der Messungen, beständig

Vertrauen. Darüber hinaus war es eine Freude, in diesem angenehmen, menschlichen Arbeitsklima mit dir, Ulli, und der Arbeitsgruppe zu arbeiten!
Im selben Zug möchte ich meinen Kolleginnen/Freundinnen/Mitbewohnerinnen Jutta Metzger und Manuela Nied danken, die mich beim Arbeiten im DESERVE Projekt, bei gemeinsamen Auslandsreisen, im Institutsalltag, und bei privaten Unternehmungen, sowie dem einen oder anderen Feierabendbierchen begleiteten. Danke für alle Diskussionen, die schöne und produktive Zusammenarbeit, euer Engagement, aber auch die gemeinsame Freizeit und den ganzen Rest!

Ohne Mitarbeit von Kollegen wäre die Realisierung meiner Messungen nicht möglich gewesen. Dabei kann ich Werner Scherer vom GPI und Martin Kohler, die mich mit viel Erfahrung und beträchtlichem Engagement auf meiner Messkampagne begleiteten, nicht genug danken. Danke für eure außergewöhnliche Unterstützung, die gute Stimmung und die tolle Zeit mit euch!
Speziell möchte ich an dieser Stelle außerdem meinen Freund Khalil Abu-Ayyash nennen. Ihn lernte ich zu Beginn des Projekts kennen. Er war ein toller Gastgeber, unterstützte unser Projekt, meine Arbeit und war in Jordanien einfach meine gute Seele. Danke.

Außerdem möchte ich den beteiligten jordanischen Kollegen und Instituten meinen Dank aussprechen:
Without the help and the commitment from my colleagues from the Ministry of Energy and Mineral Resources, who collaborated and supported this project beyond all expectations, this work could not have been achieved. My sincere thanks to Dr. Ali Sawarieh who provided staff and expertise, paving the way to conduct measurements in the Dead Sea area. Dr. Mahmoud Al-Qaryouti, head of the Seismology department, collaborated with us from the very beginning of the project, providing exceptional support and advice, establishing essential contacts for collaborations. Furthermore, I'd like to mention Mr. Alaa Muja-

hed, Manager of Panoramic Complex, who was open to the project and always hospitable, providing the grounds of the Panoramic Complex for our measurements.

Neben diversen Auslandsaufenthalten verbrachte ich auch viel schöne Zeit am Institut. Danke an Rowell, mit dem ich mir ein Büro teilte, und mit dem man auch außerhalb des Instituts eine gute Zeit hatte. Danke auch an Gabi und Rosi, die sich im Hintergrund darum kümmern, dass nichts klemmt und der Kaffeekonsum bestens dokumentiert ist (die Krankenkasse dankt). Und danke für das wirklich außergewöhnlich tolle soziale Umfeld am Institut, im besonderen der Kaffeerunde 2.0!

Meiner Familie und meinen Freunden außerhalb des Instituts sende ich hiermit ein Lebenszeichen.
Ich möchte euch danken, für mein wirklich schönes, durch euch geprägtes Leben abseits der Arbeit. Danke für gemeinsame Unternehmungen, Feiern, und Urlaube! Danke Nadine, danke Gabriel, dass ihr mir immer noch so wichtige Freunde seid. Und danke, Christoph. Trotz der mittlerweile großen Distanz warst du nie weit weg.
Außerdem eine dicke Umarmung an meine Eltern und meinen Bruder Johannes. Danke für eure Liebe und Unterstützung.
Zu guter Letzt möchte ich meinen Freund Thom nennen. Dabei ist kaum nötig zu erwähnen, dass er mich auch bei meiner Arbeit immer unterstützte. Thom, wir haben uns immer wieder zusammengerauft, du bist mein Ausgleich. Ich freue mich wieder auf mehr Zeit miteinander.

Karlsruhe, Juni 2016
Friederike Lott

Wissenschaftliche Berichte des Instituts für Meteorologie und Klimaforschung des Karlsruher Instituts für Technologie (0179-5619)

Bisher erschienen:

Nr. 1: *Fiedler, F. / Prenosil, T.*
Das MESOKLIP-Experiment. (Mesoskaliges Klimaprogramm im Oberrheintal). August 1980

Nr. 2: *Tangermann-Dlugi, G.*
Numerische Simulationen atmosphärischer Grenzschichtströmungen über langgestreckten mesoskaligen Hügelketten bei neutraler thermischer Schichtung. August 1982

Nr. 3: *Witte, N.*
Ein numerisches Modell des Wärmehaushalts fließender Gewässer unter Berücksichtigung thermischer Eingriffe. Dezember 1982

Nr. 4: *Fiedler, F. / Höschele, K. (Hrsg.)*
Prof. Dr. Max Diem zum 70. Geburtstag. Februar 1983 (vergriffen)

Nr. 5: *Adrian, G.*
Ein Initialisierungsverfahren für numerische mesoskalige Strömungsmodelle. Juli 1985

Nr. 6: *Dorwarth, G.*
Numerische Berechnung des Druckwiderstandes typischer Geländeformen. Januar 1986

Nr. 7: *Vogel, B.; Adrian, G. / Fiedler, F.*
MESOKLIP-Analysen der meteorologischen Beobachtungen von mesoskaligen Phänomenen im Oberrheingraben. November 1987

Nr. 8: *Hugelmann, C.-P.*
Differenzenverfahren zur Behandlung der Advektion. Februar 1988

Nr. 9: *Hafner, T.*
Experimentelle Untersuchung zum Druckwiderstand der Alpen. April 1988

Nr. 10: *Corsmeier, U.*
Analyse turbulenter Bewegungsvorgänge in der maritimen atmosphärischen Grenzschicht. Mai 1988

Nr. 11: *Walk, O. / Wieringa, J.(eds)*
Tsumeb Studies of the Tropical Boundary-Layer Climate. Juli 1988

Nr. 12: *Degrazia, G. A.*
Anwendung von Ähnlichkeitsverfahren auf die turbulente Diffusion in der konvektiven und stabilen Grenzschicht. Januar 1989

Nr. 13: *Schädler, G.*
Numerische Simulationen zur Wechselwirkung zwischen Landoberflächen und atmophärischer Grenzschicht. November 1990

Nr. 14: *Heldt, K.*
Untersuchungen zur Überströmung eines mikroskaligen Hindernisses in der Atmosphäre. Juli 1991

Nr. 15: *Vogel, H.*
Verteilungen reaktiver Luftbeimengungen im Lee einer Stadt – Numerische Untersuchungen der relevanten Prozesse. Juli 1991

Nr. 16: *Höschele, K.(ed.)*
Planning Applications of Urban and Building Climatology – Proceedings of the IFHP / CIB-Symposium Berlin, October 14-15, 1991. März 1992

Nr. 17: *Frank, H. P.*
Grenzschichtstruktur in Fronten. März 1992

Nr. 18: *Müller, A.*
Parallelisierung numerischer Verfahren zur Beschreibung von Ausbreitungs- und chemischen Umwandlungsprozessen in der atmosphärischen Grenzschicht. Februar 1996

Nr. 19: *Lenz, C.-J.*
Energieumsetzungen an der Erdoberfläche in gegliedertem Gelände. Juni 1996

Nr. 20: *Schwartz, A.*
Numerische Simulationen zur Massenbilanz chemisch reaktiver Substanzen im mesoskaligen Bereich. November 1996

Nr. 21: *Beheng, K. D.*
Professor Dr. Franz Fiedler zum 60. Geburtstag. Januar 1998

Nr. 22: *Niemann, V.*
Numerische Simulation turbulenter Scherströmungen mit einem Kaskadenmodell. April 1998

Nr. 23: *Koßmann, M.*
Einfluß orographisch induzierter Transportprozesse auf die Struktur der atmosphärischen Grenzschicht und die Verteilung von Spurengasen. April 1998

Nr. 24: *Baldauf, M.*
Die effektive Rauhigkeit über komplexem Gelände – Ein Störungstheoretischer Ansatz. Juni 1998

Nr. 25: *Noppel, H.*
Untersuchung des vertikalen Wärmetransports durch die Hangwindzirkulation auf regionaler Skala. Dezember 1999

Nr. 26: *Kuntze, K.*
Vertikaler Austausch und chemische Umwandlung von Spurenstoffen über topographisch gegliedertem Gelände. Oktober 2001

Nr. 27: *Wilms-Grabe, W.*
Vierdimensionale Datenassimilation als Methode zur Kopplung zweier verschiedenskaliger meteorologischer Modellsysteme. Oktober 2001

Nr. 28: *Grabe, F.*
Simulation der Wechselwirkung zwischen Atmosphäre, Vegetation und Erdoberfläche bei Verwendung unterschiedlicher Parametrisierungsansätze. Januar 2002

Nr. 29: *Riemer, N.*
Numerische Simulationen zur Wirkung des Aerosols auf die troposphärische Chemie und die Sichtweite. Mai 2002

Nr. 30: *Braun, F. J.*
Mesoskalige Modellierung der Bodenhydrologie. Dezember 2002

Nr. 31: *Kunz, M.*
Simulation von Starkniederschlägen mit langer Andauer über Mittelgebirgen. März 2003

Nr. 32: *Bäumer, D.*
Transport und chemische Umwandlung von Luftschadstoffen im Nahbereich von Autobahnen – numerische Simulationen. Juni 2003

Nr. 33: *Barthlott, C.*
Kohärente Wirbelstrukturen in der atmosphärischen Grenzschicht. Juni 2003

Nr. 34: *Wieser, A.*
Messung turbulenter Spurengasflüsse vom Flugzeug aus.
Januar 2005

Nr. 35: *Blahak, U.*
Analyse des Extinktionseffektes bei Niederschlagsmessungen mit einem C-Band Radar anhand von Simulation und Messung.
Februar 2005

Nr. 36: *Bertram, I.*
Bestimmung der Wasser- und Eismasse hochreichender konvektiver Wolken anhand von Radardaten, Modellergebnissen und konzeptioneller Betrachtungen. Mai 2005

Nr. 37: *Schmoeckel, J.*
Orographischer Einfluss auf die Strömung abgeleitet aus Sturmschäden im Schwarzwald während des Orkans „Lothar". Mai 2006

Nr. 38: *Schmitt, C.*
Interannual Variability in Antarctic Sea Ice Motion: Interannuelle Variabilität antarktischer Meereis-Drift. Mai 2006

Nr. 39: *Hasel, M.*
Strukturmerkmale und Modelldarstellung der Konvektion über Mittelgebirgen. Juli 2006

Ab Band 40 erscheinen die Wissenschaftlichen Berichte des Instituts für Meteorologie und Klimaforschung bei KIT Scientific Publishing (ISSN 0179-5619).
Die Bände sind unter www.ksp.kit.edu als PDF frei verfügbar oder als Druckausgabe bestellbar.

Nr. 40: *Lux, R.*
Modellsimulationen zur Strömungsverstärkung von orographischen Grundstrukturen bei Sturmsituationen
ISBN 978-3-86644-140-8

Nr. 41: *Straub, W.*
Der Einfluss von Gebirgswellen auf die Initiierung und Entwicklung konvektiver Wolken
ISBN 978-3-86644-226-9

Nr. 42: *Meißner, C.*
High-resolution sensitivity studies with the regional climate model COSMO-CLM
ISBN 978-3-86644-228-3

Nr. 43: *Höpfner, M.*
Charakterisierung polarer stratosphärischer Wolken mittels hochauflösender Infrarotspektroskopie
ISBN 978-3-86644-294-8

Nr. 44: *Rings, J.*
Monitoring the water content evolution of dikes
ISBN 978-3-86644-321-1

Nr. 45: *Riemer, M.*
Außertropische Umwandlung tropischer Wirbelstürme: Einfluss auf das Strömungsmuster in den mittleren Breiten
ISBN 978-3-86644-766-0

Nr. 46: *Anwender, D.*
Extratropical Transition in the Ensemble Prediction System of the ECMWF: Case Studies and Experiments
ISBN 978-3-86644-767-7

Nr. 47: *Rinke, R.*
Parametrisierung des Auswaschens von Aerosolpartikeln durch Niederschlag
ISBN 978-3-86644-768-4

Nr. 48: *Stanelle, T.*
Wechselwirkungen von Mineralstaubpartikeln mit thermodynamischen und dynamischen Prozessen in der Atmosphäre über Westafrika
ISBN 978-3-86644-769-1

Nr. 49: *Peters, T.*
Ableitung einer Beziehung zwischen der Radarreflektivität, der Niederschlagsrate und weiteren aus Radardaten abgeleiteten Parametern unter Verwendung von Methoden der multivariaten Statistik
ISBN 978-3-86644-323-5

Nr. 50: *Khodayar Pardo, S.*
High-resolution analysis of the initiation of deep convection forced by boundary-layer processes
ISBN 978-3-86644-770-7

Nr. 51: *Träumner, K.*
Einmischprozesse am Oberrand der konvektiven atmosphärischen Grenzschicht
ISBN 978-3-86644-771-4

Nr. 52: *Schwendike, J.*
Convection in an African Easterly Wave over West Africa and the Eastern Atlantic: A Model Case Study of Hurricane Helene (2006) and its Interaction with the Saharan Air Layer
ISBN 978-3-86644-772-1

Nr. 53: *Lundgren, K.*
Direct Radiative Effects of Sea Salt on the Regional Scale
ISBN 978-3-86644-773-8

Nr. 54: *Sasse, R.*
Analyse des regionalen atmosphärischen Wasserhaushalts unter Verwendung von COSMO-Simulationen und GPS-Beobachtungen
ISBN 978-3-86644-774-5

Nr. 55: *Grenzhäuser, J.*
Entwicklung neuartiger Mess- und Auswertungsstrategien für ein scannendes Wolkenradar und deren Anwendungsbereiche
ISBN 978-3-86644-775-2

Nr. 56: *Grams, C.*
Quantification of the downstream impact of extratropical transition for Typhoon Jangmi and other case studies
ISBN 978-3-86644-776-9

Nr. 57: *Keller, J.*
Diagnosing the Downstream Impact of Extratropical Transition Using Multimodel Operational Ensemble Prediction Systems
ISBN 978-3-86644-984-8

Nr. 58: *Mohr, S.*
Änderung des Gewitter- und Hagelpotentials im Klimawandel
ISBN 978-3-86644-994-7

Nr. 59: *Puskeiler, M.*
Radarbasierte Analyse der Hagelgefährdung in Deutschland
ISBN 978-3-7315-0028-5

Nr. 60: *Zeng, Y.*
Efficient Radar Forward Operator for Operational Data Assimilation within the COSMO-model
ISBN 978-3-7315-0128-2

Nr. 61: *Bangert, M. J.*
Interaction of Aerosol, Clouds, and Radiation on the Regional Scale
ISBN 978-3-7315-0123-7

Nr. 62: *Jerger, D.*
Radar Forward Operator for Verification of Cloud Resolving Simulations within the COSMO Model
ISBN 978-3-7315-0172-5

Nr. 63: *Maurer, V.*
Vorhersagbarkeit konvektiver Niederschläge : Hochauflösende Ensemblesimulationen für Westafrika
ISBN 978-3-7315-0189-3

Nr. 64: *Stawiarski, C.*
Optimizing Dual-Doppler Lidar Measurements of Surface Layer Coherent Structures with Large-Eddy Simulations
ISBN 978-3-7315-0197-8

Nr. 65: *Mahlke, H.*
Mechanismen der Auslösung hochreichender Konvektion im südwestdeutschen Mittelgebirgsraum
ISBN 978-3-7315-0203-6

Nr. 66: *Fosser, G.*
Precipitation statistics from regional climate model at resolutions relevant for soil erosion
ISBN 978-3-7315-0227-2

Nr. 67: *Adler, B.*
Boundary-Layer Processes Producing Mesoscale Water-Vapour Variability over a Mountainous Island
ISBN 978-3-7315-0247-0

Nr. 68: *Kraut, I.*
Separating the Aerosol Effect in Case of a „Medicane"
ISBN 978-3-7315-0405-4

Nr. 69: *Breil, M.*
Einfluss der Boden-Vegetation-Atmosphären Wechselwirkungen auf die dekadische Vorhersagbarkeit des Westafrikanischen Monsuns
ISBN 978-3-7315-0420-7

Nr. 70: *Lott, F. F.*
Wind Systems in the Dead Sea and Footprints in Seismic Records
ISBN 978-3-7315-0596-9